AF745559

Rotating Machinery

Edited by Getu Hailu

Published in London, United Kingdom

IntechOpen

Supporting open minds since 2005

Rotating Machinery
http://dx.doi.org/10.5772/intechopen.77579
Edited by Getu Hailu

Contributors
Valder Steffen Jr, Aldemir Ap. Cavalini Jr, Fangyuan Lou, Rodrigo Serrano, José Moreira, Ana Letícia Pilz Castro, Edna Viana, Maria Pinto, Calos Martinez, Akanksha Singh, Yan Li, Getu Hailu

© The Editor(s) and the Author(s) 2020
The rights of the editor(s) and the author(s) have been asserted in accordance with the Copyright, Designs and Patents Act 1988. All rights to the book as a whole are reserved by INTECHOPEN LIMITED. The book as a whole (compilation) cannot be reproduced, distributed or used for commercial or non-commercial purposes without INTECHOPEN LIMITED's written permission. Enquiries concerning the use of the book should be directed to INTECHOPEN LIMITED rights and permissions department (permissions@intechopen.com).
Violations are liable to prosecution under the governing Copyright Law.

(cc) BY

Individual chapters of this publication are distributed under the terms of the Creative Commons Attribution 3.0 Unported License which permits commercial use, distribution and reproduction of the individual chapters, provided the original author(s) and source publication are appropriately acknowledged. If so indicated, certain images may not be included under the Creative Commons license. In such cases users will need to obtain permission from the license holder to reproduce the material. More details and guidelines concerning content reuse and adaptation can be found at http://www.intechopen.com/copyright-policy.html.

Notice
Statements and opinions expressed in the chapters are these of the individual contributors and not necessarily those of the editors or publisher. No responsibility is accepted for the accuracy of information contained in the published chapters. The publisher assumes no responsibility for any damage or injury to persons or property arising out of the use of any materials, instructions, methods or ideas contained in the book.

First published in London, United Kingdom, 2020 by IntechOpen
IntechOpen is the global imprint of INTECHOPEN LIMITED, registered in England and Wales, registration number: 11086078, 7th floor, 10 Lower Thames Street, London, EC3R 6AF, United Kingdom
Printed in Croatia

British Library Cataloguing-in-Publication Data
A catalogue record for this book is available from the British Library

Additional hard and PDF copies can be obtained from orders@intechopen.com

Rotating Machinery
Edited by Getu Hailu
p. cm.
Print ISBN 978-1-78984-025-4
Online ISBN 978-1-78984-026-1
eBook (PDF) ISBN 978-1-83968-157-8

For EU product safety concerns:
IN TECH d.o.o., Prolaz Marije Krucifikse Kozulić 3, 51000 Rijeka, Croatia, info@intechopen.com or visit our website at intechopen.com.

We are IntechOpen,
the world's leading publisher of
Open Access books

Built by scientists, for scientists

4,500+
Open access books available

118,000+
International authors and editors

130M+
Downloads

151
Countries delivered to

Our authors are among the
Top 1%
most cited scientists

12.2%
Contributors from top 500 universities

WEB OF SCIENCE™

Selection of our books indexed in the Book Citation Index
in Web of Science™ Core Collection (BKCI)

Interested in publishing with us?
Contact book.department@intechopen.com

Numbers displayed above are based on latest data collected.
For more information visit www.intechopen.com

Meet the editor

Dr Getu Hailu is currently an Assistant Professor of mechanical engineering at the University of Alaska Anchorage. He has more than 20 years of experience in research and teaching. His area of expertise is thermofluids, including thermal storages, building integrated photovoltaic thermal systems, solar energy, and air source heat pump application in cold regions. His teaching interests include courses in the thermofluids and renewable energy areas including turbomachinery, both at the graduate and undergraduate level. He has authored/coauthored more than 40 refereed publications. He is a member of ASME, ASHRAE, and IBPSA.

Contents

Preface

The topic of turbomachinery is vast and includes all rotating machines that transfer energy between a rotor and a fluid. These include wind turbines, hydro turbines, pumps, fans, and compressors. There is constant improvement in the design, testing, and failure diagnosis of turbomachinery. This book covers some selected topics of turbomachinery, such as advanced measurement techniques, abrasive wear of parts of turbomachinery, and uncertainty analysis. Modern measurement techniques for turbomachinery application including temperature, pressure, velocity, and strain/stress measurements are included. A chapter is dedicated to methods of evaluation of abrasive wear of pump impellers.

Currently much attention is being given to renewable energy sources, and many countries have set a goal of reaching net zero energy in the near future. Wind turbines are among renewable energy technologies that are intended to help reach the goal of net zero energy in buildings. There are two types of wind turbines that are widely used: horizontal axis and vertical axis wind turbines (HAWTs and VAWTs). VAWTs have fewer parts when compared to HAWTs. VAWTs do not require yaw mechanisms for orienting the wind turbines towards the wind direction; the gearbox and the generator are located on the ground; and the tower does not have to be as strong as those used for HAWTs. Because of these reasons, VAWTs are more practical for residential and small-scale applications. A chapter is dedicated to VAWTs.

Getu Hailu
University of Alaska,
USA

Section 1

Introduction

Chapter 1

Introductory Chapter: Rotating Machinery

Getu Hailu

1. Introduction

Rotating machinery or turbomachinery is a machine with a rotating component that transfers energy to a fluid or vice versa. Consequently, in a turbomachine there is energy transfer between the fluid and the rotor through dynamic interaction. Generally, if the energy transfer is from the rotor to the fluid, it is either a pump or fan. If the energy transfer is from the fluid to the rotor, then the machine is called turbine.

We come across a turbomachine in everyday life. In fact, we come to use a turbomachine at least once in a day. When we dry our hair with a hair dryer, we are using a turbomachine. A hair dryer blows room temperature air over nichrome (alloy of nickel, chromium, and often iron) hot coils speeding up water evaporation. The important component of the hair dryer is the one that blows air (aka fan). This component is a turbomachine. Another commonly used household machine is the clothes washer. The washing machines need to drain the used dirty water and replace it with fresh water. To do so an important component of a washing machine is a pump that is used to remove the dirty water and supply fresh water. This pump is a turbomachine.

If you own a car, you know how important it is to maintain an optimum operating temperature of your car. A water pump (hydrodynamic pump) is essential to your car's operation. The pump ensures that the coolant keeps circulating through the engine block, hoses, and radiator and maintains an optimum operating temperature.

Another everyday example is a kitchen vent. Fans inside the kitchen vent that pull the fumes in and push them via ductwork to the outside or through filters (that remove odors) and vent them back into the room are turbomachines.

Besides a washing machine, a dishwasher, or a kitchen vent, when we are writing on our laptop/desktop, when we turn on the bathroom vent, and when we turn on a desktop fan or a ceiling fan, we are using turbomachines.

All these components, the little fan in our computer which helps maintain the temperature of our computer or the ceiling fan which provides the thermal comfort needed in summer times, have certain types of geometry and shapes. You have probably noticed the difference in shapes and the number of blades between a windmill and modern wind turbine. These shapes and numbers are a result of careful analysis of fluid flow or air flow through these machines.

2. Advances in rotating machinery research

As turbomachines are key machines used in power generation and energy conversion, recent research focus has been on improving the aero-thermal performance of these machines and their efficiencies. In the aviation industry, research

IntechOpen

has focused on reducing environmental impact and fuel consumption. Most of the current research in rotating machinery is concentrated mainly in the following areas [1–7]:

- Fluid dynamics of turbomachinery (numerical simulation, theoretical model, and experimental measurement)
- Noise reduction methods and noise testing technologies in fans, compressors, pumps, and wind turbines
- Optimization methods
- Turbomachinery materials performance enhancement

Although turbomachines are one of the most widely used machines, the two main areas of applications are power generation and propulsion. Current high-level research topics related to power generation turbines include numerical modeling of two-phase flow fields, and research efforts are aimed at reducing erosion (cavitation erosion and liquid drop erosion) in these machines [8, 9, 10]. Computational fluid dynamics (CFD) is being increasingly used to assess blade design with respect to aero-elastic instability for improving component life span and efficiency [11]. CFD analysis is used to minimize cavitation effects in hydraulic turbines, thus leading to better performance, efficiency, and cost savings. Identification of alternative fuels emission reduction is another area of research as environmental regulations get stricter. Research focus includes studying the combustion characteristics of alternative fuels (ethanol, palm methyl ester (PME), dimethyl ester (DME), hydrogen/syngas, and biofuels [12–17]).

3. Concluding remarks

Turbomachines are essential for they have key applications including power generation. Consequently, research and development is driven by the relevant industry. Research focus thus is on increased efficiency (geared toward cost savings), reliability (increasing life span of a component), and sustainability (driven by stricter regulations). CFD has emerged to be important in designing and analyzing turbomachinery components.

Author details

Getu Hailu
Mechanical Engineering, University of Alaska Anchorage, United States

*Address all correspondence to: ghailu@alaska.edu

IntechOpen

© 2020 The Author(s). Licensee IntechOpen. This chapter is distributed under the terms of the Creative Commons Attribution License (http://creativecommons.org/licenses/by/3.0), which permits unrestricted use, distribution, and reproduction in any medium, provided the original work is properly cited.

References

[1] Srinivas G, Raghunandana K, Satish Shenoy B. Recent developments in turbomachinery component materials and manufacturing challenges for aero engine applications. Materials Science and Engineering. 2018;**314**:012012

[2] Chandrasekhar U, Yang L-J, Gowthaman S, editors. Innovative Design, Analysis and Development Practices in Aerospace and Automotive Engineering (I-DAD 2018). Springer: Nature Singapore Pte Ltd.; 2019

[3] Kelkar R, Andreaco A, Ott E, Groh J. Alloy 718: Laser powder bed additive manufacturing for turbine applications. In: Ott E et al, editors. Proceedings of the 9th International Symposium on Superalloy 718 & Derivatives: Energy, Aerospace, and Industrial Applications. The Minerals, Metals & Materials Series. Cham: Springer; 2018

[4] Liu R, Wang Z, Sparks T, Liou F, Newkirk J. Aerospace component repair with additive manufacturing. In: Laser Additive Manufacturing. Elsevier; 2017

[5] Amirante D, Hills NJ, Barnes CJ. A moving mesh algorithm for aero-thermo-mechanical modelling in turbomachinery. International Journal for Numerical Methods in Fluids. 2012;**70**(9)

[6] Bucchi A, Xing JT, and Gaudenzi P. Numerical solution for a fluid-active structure interaction. In: 19th International Conference on Adaptive Structures and Technologies 2008, ICAST 2008; 2008

[7] Wang Y, Cai X, Ma X, Tan J, Liu D, Ren D. Meshfree simulation of flow around airfoil using different turbulent models. Progress in Computational Fluid Dynamics: An International Journal. 2017;**17**(2)

[8] Xie DM, Shi Y, Li WF, Hou YM, Yu XG, and Qin HS. Numerical simulation of wet steam two-phase flow in the last-stage stationary blade of super-critical steam turbine. In: DRPT 2011-2011 4th International Conference on Electric Utility Deregulation and Restructuring and Power Technologies. 2011

[9] Li N, Zhou Q, Chen X, Xu T, Hui S, Zhang D. Liquid drop impact on solid surface with application to water drop erosion on turbine blades, part I: Nonlinear wave model and solution of one-dimensional impact. International Journal of Mechanical Sciences. 2008;**50**(10-11):1526-1542

[10] Zhou Q, Li N, Chen X, Xu T, Hui S, Zhang D. Liquid drop impact on solid surface with application to water drop erosion on turbine blades, part II: Axisymmetric solution and erosion analysis. International Journal of Mechanical Sciences. 2008;**50**(10-11):1543-1558

[11] Rice T, Bell D, Singh G. Identification of the stability margin between safe operation and the onset of blade flutter. Journal of Turbomachinery. 2009;**131**(1):011009

[12] Haas FM, Chaos M, Dryer FL. Low and intermediate temperature oxidation of ethanol and ethanol-PRF blends: An experimental and modeling study. Combustion and Flame. 2009;**156**(12):2346-2350

[13] Hashimoto N, Ozawa Y, Mori N, Yuri I, Hisamatsu T. Fundamental combustion characteristics of palm methyl ester (PME) as alternative fuel for gas turbines. Fuel. 2008

[14] Lee MC, Bin Seo S, Chung JH, Joo YJ, Ahn DH. Industrial gas turbine combustion performance test of DME to use as an alternative fuel for power generation. Fuel. 2009;**87**(15-16):3373-3378

[15] Miyama N, Inaba K, Yamamoto M. Numerical simulation of tip leakage vortex effect on hydrogen-combustion flow around 3D turbine blade. Journal of Thermal Science. 2008;**17**(2):186-192

[16] Chaos M, Dryer FL. Syngas combustion kinetics and applications. Combustion Science and Technology. 2008;**180**(6):1053-1096

[17] Gupta KK, Rehman A, Sarviya RM. Bio-fuels for the gas turbine: A review. Renewable and Sustainable Energy Reviews. 2010;**14**(9):2946-2955

Section 2

Measurement, Control and Uncertainty Analysis of Turbo Machines

Chapter 2

Uncertainty Analysis Techniques Applied to Rotating Machines

Fabian Andres Lara-Molina, Arinan De Piemonte Dourado, Aldemir Ap. Cavalini Jr and Valder Steffen Jr

Abstract

This chapter presents the modeling procedure, numerical application, and experimental validation of uncertain quantification techniques applied to flexible rotor systems. The uncertainty modeling is based both on the stochastic and fuzzy approaches. The stochastic approach creates a representative model for the flexible rotor system by using the stochastic finite element method. In this case, the uncertain parameters of the rotating machine are characterized by homogeneous Gaussian random fields expressed in a spectral form by using the Karhunen-Loève (KL) expansion. The fuzzy approach uses the fuzzy finite element method, which is based on the α-level optimization. A comparative study regarding the numerical and experimental results obtained from a flexible rotor test rig is analyzed for the stochastic and fuzzy approaches.

Keywords: rotordynamics, uncertainty, fuzziness, randomness, experimental validation

1. Introduction

Rotating machines are unavoidably subjected to uncertainties that affect their parameters and, consequently, their dynamic behavior. Thus, mathematical models that encompass variability and randomness are required for the analysis and design of rotating machines instead of using deterministic models.

Uncertain dynamic responses of flexible rotors have been analyzed by applying two main approaches, namely, stochastic and fuzzy. Thus, the uncertainty analysis has been applied in flexible rotors by using the polynomial chaos theory [1], as modeled by considering Gaussian homogeneous stochastic fields discretized by Karhunen-Loève expansion [2] or through the fuzzy approach [3, 4]. These methods are well-established tools that may present limitations and drawbacks depending on the application conveyed.

In this context, this chapter presents two different approaches to model uncertain parameters and to simulate the uncertain dynamic responses of rotating machines. In this way, the stochastic and fuzzy approaches are applied to different parameters of a flexible rotor. The procedure used to obtain the stochastic model of the rotor is based on the stochastic finite element method. Moreover, the fuzzy finite element model of the rotor system is formulated according to the fuzzy approach. Then, the corresponding numerical method used to compute the fuzzy dynamic responses of the rotating machine is described. A comparative study

IntechOpen

between the stochastic and fuzzy approaches along with the validation of the obtained results by using experimental data is presented.

2. Rotor system model

The deterministic model of a flexible rotor based on the finite element method (FE model) is obtained in this section by following the formulation previously presented in [5]. The rotor system is composed of a flexible shaft, rigid discs, and bearings. **Figure 1** shows the finite element used to represent the shaft. In this case, the finite element has two nodes and four degrees of freedom (DOFs) per node. The DOFs are associated with the nodal displacements along the x and z directions (defined by u and w, respectively) and the rotations around the x and z directions ($\theta = \partial w/\partial y$ and $\psi = \partial u/\partial y$, respectively).

In this contribution, the FE model of the shaft was obtained based on the Euler-Bernoulli and Timoshenko beam theories. The displacement field along the finite element is represented by a cubic interpolation function. Therefore, $\mathbf{u}(y, t) = \mathbf{N}(y)$ $\mathbf{u}_e(t)$, where $\mathbf{N}(y)$ is a matrix containing shape interpolation functions and $\mathbf{u}_e(t) = [u_i\ w_i\ \theta_i\ \psi_i]^T$ $(i = 1, 2)$ is the vector of DOFs.

The strain and kinetic energies of the shaft finite element are defined according to analytical equations derived from the variational principle. Therefore, the mass and stiffness elementary matrices of the shaft are given by

$$
\begin{aligned}
\mathbf{M}_s^e &= \int_{y=0}^{L} \mathbf{N}_{mi}^T(y)\mathbf{N}_{mi}(y)\,dy \\
\mathbf{G}_s^e &= \int_{y=0}^{L} \mathbf{N}_g^T(y)\mathbf{N}_g(y)\mathrm{d}y \\
\mathbf{K}_s^e &= \int_{y=0}^{L} \mathbf{B}^T(y)\mathbf{E}\mathbf{B}(y)\mathrm{d}y
\end{aligned}
\tag{1}
$$

where $\mathbf{M}_s^e$ $(N_e \times N_e)$ is the elementary mass matrix of the shaft element, $\mathbf{G}_s^e$ $(N_e \times N_e)$ is the gyroscopic matrix, $\mathbf{K}_s^e$ $(N_e \times N_e)$ is the stiffness matrix, and $\mathbf{E}$ is the isotropic matrix that contains the elastic properties of the material. $\mathbf{B}(y)$ is the matrix composed of differential operators that characterize the strain–displacement

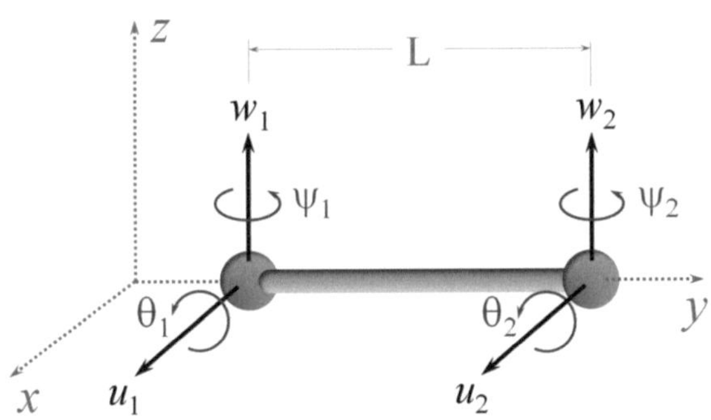

Figure 1.
Finite element of the shaft [2].

relationship. $\mathbf{N}_{mi}{}^T$ and $\mathbf{N}_g{}^T$ represent the shape interpolation functions associated with the mass and inertia matrices, respectively. $N_e = 8$ is the number of DOFs considered in the shaft finite element.

Rigid discs are introduced in the global FE model of the shaft by considering their corresponding kinetic energy. Thus, $\mathbf{M}_d^e$ $(N_d \times N_d)$ and $\mathbf{G}_d^e$ $(N_d \times N_d)$ are the mass and gyroscopic matrices associated with each disc (N_d = y is the number of DOFs considered for the disc). Moreover, the bearings are modeled by using linear stiffness and damping coefficients that are introduced conveniently in the stiffness and damping matrices of the shaft FE model, respectively [6].

Eq. (2) presents the differential equation that characterizes the dynamic behavior of rotating machines (FE model with N DOFs), which is obtained by assembling the elementary finite element matrices of the shaft:

$$\mathbf{M}\ddot{\mathbf{q}}(t) + [\mathbf{C} + \Omega\mathbf{G}]\dot{\mathbf{q}}(t) + \mathbf{K}\mathbf{q}(t) = \mathbf{F}(t) \tag{2}$$

where $\mathbf{M} = \mathbf{M}_s + \mathbf{M}_d$ $(N \times N)$ and $\mathbf{K}=\mathbf{K}_s + \mathbf{K}_b$ $(N \times N)$ are the global mass and stiffness matrices of the rotor model, respectively. $\mathbf{K}_b$ is the matrix containing the stiffness coefficients of the bearings. $\mathbf{C} = \mathbf{C}_b + \mathbf{C}_p$ $(N \times N)$ is the damping matrix that considers the damping coefficients of the bearings (matrix $\mathbf{C}_b$) and the proportional damping $\mathbf{C}_p = \alpha\mathbf{M} + \beta\mathbf{K}$ (α and β are the so-called proportional coefficients). $\mathbf{G} = \mathbf{G}_s + \mathbf{G}_d$ (NxN) is the gyroscopic matrix. $\mathbf{q}(t)$ $(N \times 1)$ and $\mathbf{F}(t)$ $(N \times 1)$ are the vectors of DOFs and external loads, respectively. Ω is the rotation speed of the shaft. More details about the formulation of the rotor FE model adopted in the present contribution can be found in [5].

3. Stochastic modeling

Among the various methods used to model uncertainties, the stochastic finite element method (SFEM) has been widely applied to complex engineering systems of industrial applications. SFEM presents well-established mathematical fundaments and suitable experimental validation [7]. Some details about the formulation of the SFEM are presented next.

3.1 Stochastic modeling of flexible shafts

The Karhunen-Loève (KL) expansion is used to model the random fields as a spectral representation. Consequently, a random field is represented as a spatial expansion of a random variable that fluctuates randomly. For instance, uncertainties affecting Young's modulus of the shaft can be evaluated by using the KL expansion. A one-dimensional random field $H(y, \theta)$ can be defined as [8]

$$H(y,\theta) = E(y) + \sum_{r=1}^{n_{KL}} \sqrt{\lambda_r} f_r(y)\xi_r(\theta) \tag{3}$$

where $f_r(y)$ and λ_r are the eigenfunctions and eigenvalues of the covariance function $C(y_1, y_2)$, respectively. n_{KL} is the number of terms used in the KL expansion.

In this work, the exponential covariance is adopted, which is defined as $C(y_1, y_2) = \mathrm{e}^{(-|y_1-y_2|/Lc)}$, where $(y_1, y_2) \in [0, L]$ and L_c represent the correlation length. $\xi_r(\theta)$ denotes the random variables that are orthonormal with respect to the functions $f_r(y)$. The KL expansion is used to model the stochastic finite element matrices of the flexible shaft, as given by Eq. (4):

$$\mathbf{M}_s^e(\theta) = \mathbf{M}_s + \sum_{r=1}^{n_{KL}} \overline{\mathbf{M}}_{sr}^e \xi_r(\theta)$$

$$\mathbf{K}_s^e(\theta) = \mathbf{K}_s + \sum_{r=1}^{n_{KL}} \overline{\mathbf{K}}_{sr}^e \xi_r(\theta) \tag{4}$$

$$\mathbf{G}_s^e(\theta) = \mathbf{G}_s + \sum_{r=1}^{n_{KL}} \overline{\mathbf{G}}_{sr}^e \xi_r(\theta)$$

where $\mathbf{M}_s$, $\mathbf{K}_s$, and $\mathbf{G}_s$ are the deterministic elementary mass, stiffness, and gyroscopic matrices of the shaft, respectively. The stochastic matrices are obtained by solving the following expressions:

$$\overline{\mathbf{M}}_{sr}^e = \int_{y=0}^{L} \sqrt{\lambda_r} f_r(y) \mathbf{N}_{mi}^T(y) \mathbf{N}_{mi}(y) dy$$

$$\overline{\mathbf{K}}_{sr}^e = \int_{y=0}^{L} \sqrt{\lambda_r} f_r(y) \mathbf{B}^T(y) \overline{\mathbf{E}} \mathbf{B}(y) dy \tag{5}$$

$$\overline{\mathbf{G}}_{sr}^e = \int_{y=0}^{L} \sqrt{\lambda_r} f_r(y) \mathbf{N}_g^T(y) \mathbf{N}_g(y) dy$$

in which $\overline{\mathbf{E}}$ is the mechanical property matrix that contains the parameters E_s, A_s, and I_s (Young's modulus, cross-sectional area, and inertia moment of the shaft, respectively).

3.2 Stochastic modeling of bearings' parameters

The uncertainties associated with bearings' stiffness and damping coefficients of rotating machines can be evaluated by using the following relations: $k(\theta) = k_o + k_o \delta_k \xi(\theta)$ and $d(\theta) = d_o + d_o \delta_d \xi(\theta)$, respectively. In this case, k_o and d_o are the mean values of the stiffness and damping coefficients of the bearings, respectively. δ_k and δ_d are the corresponding dispersion levels. $\xi(\theta)$ represents the stochastic distribution. The stochastic model of the rotor is solved by using the Monte Carlo simulation (MCS) in combination with Latin hypercube sampling [9].

3.3 Numerical results

In this section, SFEM is applied to the FE model as given by **Figure 2**. The rotating machine is composed of a horizontal flexible shaft discretized into 20 Euler-Bernoulli's beam elements, three asymmetric bearings (B_1, B_2, and B_3), and two rigid discs (D_1 and D_2). The physical and geometrical characteristics used in the FE model of the rotor system are given in [2].

In this case, the uncertain random fields associated with Young's modulus of the shaft are modeled as homogeneous Gaussian stochastic fields, which are represented in the spectral form by using the Karhunen-Loève expansion. The uncertainty variables associated with the stiffness and damping coefficients of the

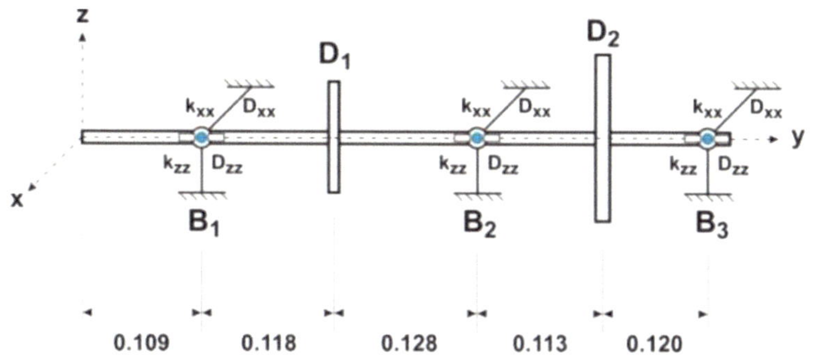

Figure 2.
Numerical model of the flexible rotor system [2].

Scenario	n_{KL}	n_s
(a)	$1 \leq n_{KL} \leq 50$	100
(b)	10	$1 \leq n_s \leq 250$

Table 1.
Parameters of convergence analysis simulation.

bearings are modeled as random variables. This modeling process considers the frequency- and time-domain vibration responses of the rotating machine in terms of their working envelopes (frequency response functions (FRFs) and orbits).

Initially, the convergence of the stochastic model is verified by changing the number of terms used in the KL expansion and the number of samples considered in MCS (n_{KL} and n_s, respectively). The convergence analysis was performed based on the root-mean-square (RMS) value as given by Eq. (6):

$$RMS = \left[\frac{1}{n_s} \sum_{j=1}^{n_s} \left| \mathbf{H}_j(\omega, \Omega, \theta) - \mathbf{H}_j(\omega, \Omega) \right|^2 \right]^{\frac{1}{2}} \tag{6}$$

where $\mathbf{H}(\omega, \Omega)$ is the FRF obtained by using the deterministic FE model of the rotor and $\mathbf{H}(\omega, \Omega, \theta)$ is the corresponding FRF of the stochastic model associated with independent realizations θ. In this case, ω is the frequency.

The deterministic and stochastic FRFs were obtained by considering the shaft at rest ($\Omega = 0$) from impacts performed along the x direction of the disc D_1 and measures obtained at the same position and direction. Two scenarios were evaluated to achieve convergence for n_{KL} and n_s, as given by **Table 1**. In both cases, the correlation length L_C was assumed as being equal to the length of the shaft elements.

Figures 3a and **b** present the upper and lower limits of the RMS envelopes obtained by considering the scenarios (a) and (b) of **Table 1**, respectively. Note that convergence is achieved for n_{KL} = 10 and n_s = 70.

Figure 4a and **b** show the FRF and orbit, respectively, obtained by using the deterministic (mean) and stochastic FE models of the rotor system. The uncertain envelopes were determined by applying a 5% dispersion level both in Young's modulus of the shaft (E_s) and in the stiffness and damping coefficients of the bearings (k_{xx}, k_{zz}, d_{xx}, and d_{zz}; see **Figure 2**). The results show the influence of the uncertain parameters on the dynamic behavior of the flexible rotor, which are highlighted by the dispersion of the uncertain envelopes around the curves of the deterministic FRF and orbit (mean model) (**Figures 3** and **4**).

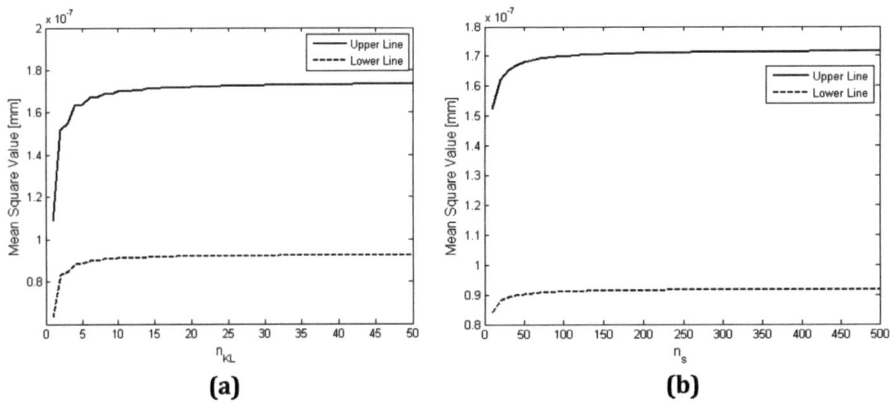

Figure 3.
Convergence simulation: (a) n_{KL} and (b) n_s [2].

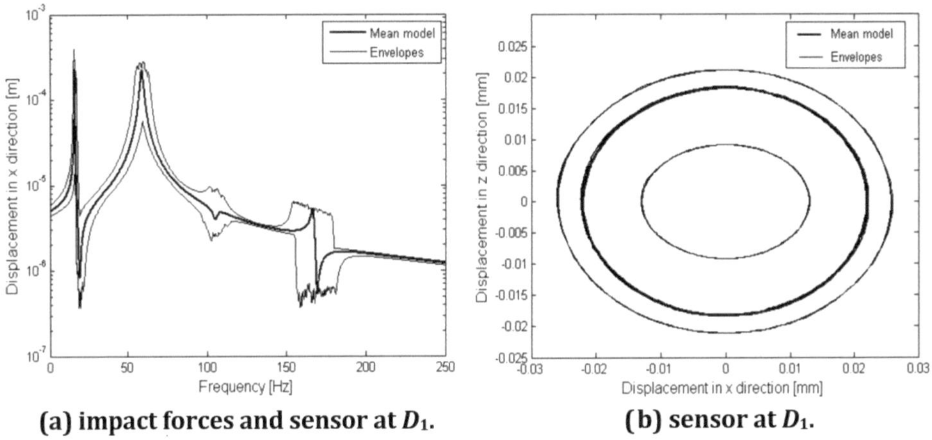

Figure 4.
Stochastic responses: (a) FRF and (b) orbits [2].

4. Fuzzy dynamic analysis

The fuzzy dynamic analysis computes the uncertain dynamic responses of rotating machines by modeling the uncertain parameters as fuzzy variables or fuzzy fields. The fuzzy dynamic analysis is based on the α-level optimization, which was introduced by [10]. In the α-level approach, an optimization problem should be solved to compute the fuzzy responses of the system as presented next.

4.1 Fuzzy variables

Figure 5 presents the definition of fuzzy sets. Considering $\mathbf{X}$ as a universal set whose elements are defined by x, subset A ($A \in \mathbf{X}$) is defined by the membership function μ_A: $\mathbf{X} \rightarrow \{0, 1\}$, where μ_A is a membership function with real value and continuous interval. Each element belongs (for $\mu_A = 1$) or does not belong to the classical set A (see **Figure 5a**). Moreover, a fuzzy set $\tilde{A}$ is defined by the membership function μ_A: $\mathbf{X} \rightarrow [0, 1]$. The membership function $\mu_A(x)$ defines how compatible the element x is with respect to the fuzzy set $\tilde{A}$. Thus, $\mu_A(x)$ close to 1 indicates high pertinence of x to $\tilde{A}$.

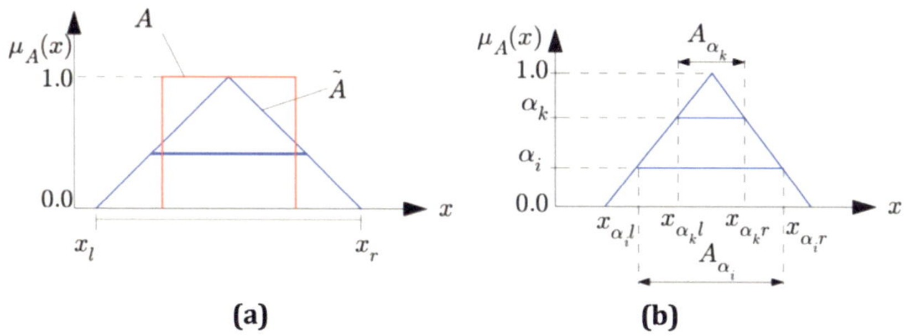

Figure 5.
Fuzzy set: (a) definition and (b) α-level representation [11].

Fuzzy variables are represented by using intervals weighted by the membership function, namely, α-levels. According to the α-level representation, $\tilde{A}$ is defined as

$$\tilde{A} = \{(x, \mu_A(x)) | x \in \mathbf{X}\} \tag{7}$$

where $0 \leq \mu_A(x) \leq 1$.

Moreover, according to **Figure 5b**

$$\tilde{A}_{\alpha_k} = \{x \in \mathbf{X}, \mu_A(x) \geq \alpha_k\} \tag{8}$$

If the fuzzy set is convex, each α-level subset $A_{\alpha k}$ corresponds to the interval $[x_{\alpha kl}, \mathrm{x}_{\alpha ku}]$, where

$$\begin{aligned} x_{\alpha_k l} &= \min\,[x \in \mathbf{X}, \mu_A(x) \geq \alpha_k] \\ x_{\alpha_k u} &= \max\,[x \in \mathbf{X}, \mu_A(x) \geq \alpha_k] \end{aligned} \tag{9}$$

4.2 Fuzzy dynamic analysis

The fuzzy dynamic analysis is a numerical method used to map a fuzzy input $\underset{\sim}{\tilde{x}}$ onto a fuzzy output $\underset{\sim}{\tilde{z}}(\tau)$ by using deterministic models, as given by Eq. (2). Thus, the fuzzy finite element method is defined by combining fuzzy parameters (uncertain information) with a deterministic model based on the classic FE method.

Figure 6 shows that the fuzzy dynamic analysis is composed of two main steps. The first step consists in discretizing the input fuzzy parameter according to the α-level representation presented in Eq. (8) and **Figure 5b**. Thus, each fuzzy parameter of the vector $\underset{\sim}{\tilde{x}} = (\tilde{x}_1, \ldots, \tilde{x}_n)$ is represented by an interval $X_{i\alpha k} = [x_{i\alpha kl}, x_{i\alpha ku}]$, where α_k $\mathbb{R}$ $[0, 1]$. Therefore, the subspace crisp $\underline{X}_{\alpha k}$ is defined as $\underline{X}_{\alpha k} = (X_{1\alpha k}, \ldots X_{n\alpha k}) \in \mathbb{R}^n$.

In the second step, an optimization problem is performed. This optimization process maximizes and minimizes the value of the output for the mapping model $M : \underline{z} = f(\underline{x})$ and over the subspace crisp at each evaluated value τ. Thus,

$$\begin{aligned} z_{\alpha_k l} &= \min_{\underline{x} \in \underline{X}\alpha_k} f(\underline{x}, \tau) \\ z_{\alpha_k u} &= \max_{\underline{x} \in \underline{X}\alpha_k} f(\underline{x}, \tau) \end{aligned} \tag{10}$$

where $z_{\alpha kl}$ and $z_{\alpha ku}$ are the lower and upper limits of the interval $z_{\alpha k} = [z_{\alpha kl}, z_{\alpha ku}]$ corresponding to the α-level α_k.

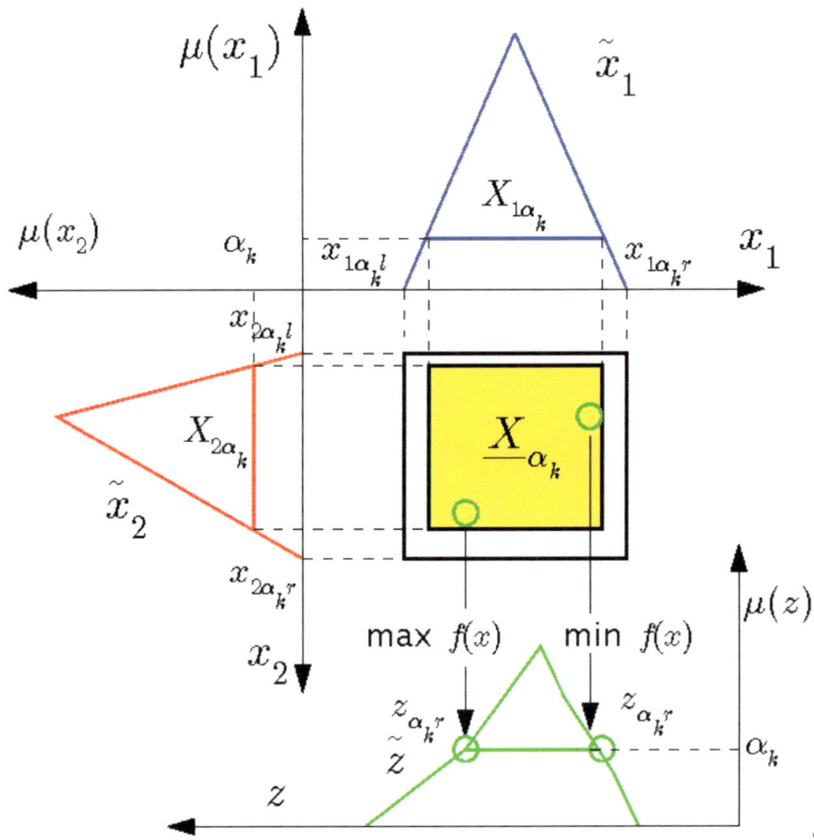

Figure 6.
α-Level optimization.

The complete set of the intervals $z_{\alpha k}$ for $\alpha_k \in [0, 1]$ forms the fuzzy resulting variable $\tilde{z}(\tau)$ evaluated at τ.

The fuzzy analysis of either a transient time-domain response or a frequency response function demands the solution of a large number of α-level optimization processes, i.e., one α-level optimization at each considered time or frequency step. In the present contribution, the optimization associated with the α-levels is solved by using the differential evolution optimization algorithm [12].

4.3 Numerical results

The numerical results for the fuzzy analysis are also obtained by using the rotor FE model presented in **Figure 2**. In this case, Young's modulus E_S of the shaft and the stiffness and damping coefficients of the bearings (B_1, B_2, and B_3) were considered as fuzzy triangular numbers (uncertain parameters). In this case, a 5% dispersion level was applied around the deterministic value of Young's modulus and a 15% dispersion level around the deterministic values of the stiffness and damping coefficients of the bearings. The fuzzy response of the rotor system was assessed at three different α-levels: 0, 0.5, and 1.0. **Figure 7a** and **b** shows the FRF and orbit obtained by applying the fuzzy uncertain analysis technique, respectively.

The fuzzy responses both on the time and frequency domains show that the fuzzy uncertainty parameters produce a significant variation of the lower and upper curves of the fuzzy envelope. Note that the results obtained in the present analysis are similar to the ones presented in **Figure 4**, for which the stochastic approach was applied.

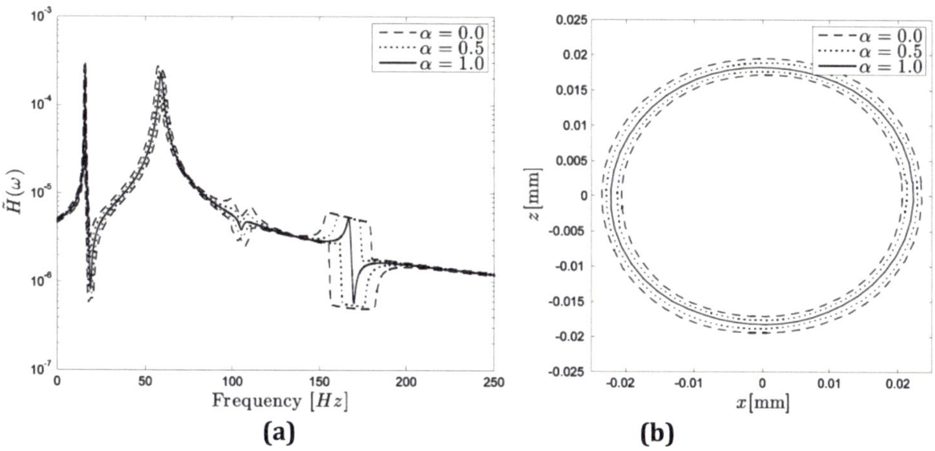

Figure 7.
Fuzzy responses: (a) FRF and (b) orbit [3].

5. Comparative study of uncertainty quantification techniques

The uncertainty analysis of dynamic systems has been previously studied by applying techniques based both on stochastic and fuzzy approaches. The fuzzy approach has demonstrated to be more appropriate in the cases of applications for which there is no knowledge regarding the stochastic process that governs the uncertainties themselves.

In the present study, the uncertainties that affect the dynamic response of a flexible rotor system are modeled by using both stochastic and fuzzy approaches. These methodologies have been compared by evaluating the dynamic responses obtained by numerical simulations regarding the frequency responses and time-domain responses. The numerical and experimental results of this section have been obtained from the flexible rotor test rig depicted in **Figure 8**.

The corresponding FE model was discretized in 33 finite elements, as given by **Figure 8b**. This rotating machine is composed of a flexible steel shaft of 860 length and 17 mm diameter (E = 205 GPa, ρ = 7850 kg/m^3, υ = 0.29); two ball bearings (B_1 and B_2); located at nodes #4 and #31, respectively; and two rigid discs D_1 (located at node #13) and D_2 (at node #23). Displacement sensors are placed at nodes #8 (S_{8X} and S_{8Z}) and #28 (S_{28X} and S_{28Z}) to measure the shaft vibration responses. An electric DC motor drives the shaft.

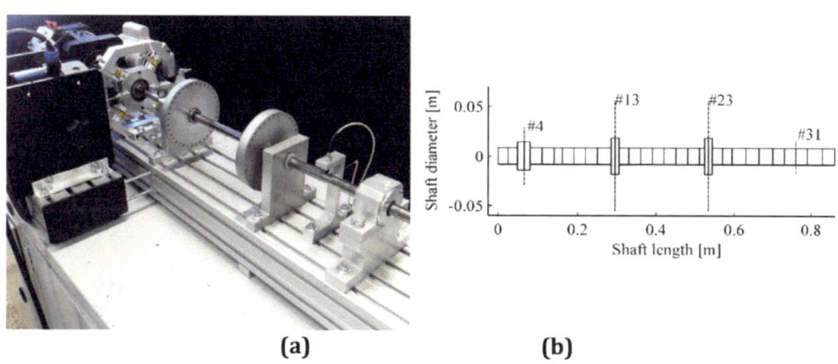

Figure 8.
Experimental rotor: (a) test rig and (b) FE model [13].

A representative FE model of the rotating machine was obtained by applying a model updating procedure. The differential evolution optimization approach was used to identify the unknown parameters of the FE model, namely, coefficients α and β (proportional damping), the stiffness and damping coefficients of the bearings, and the angular stiffness k_{ROT} introduced by the coupling between the electric motor and the shaft (orthogonal to plane XZ at node #1). Further information about the model updating procedure can be found in [8].

Figure 9 shows the simulated Bode diagram obtained by using the parameters identified by the considered optimization procedure. The experimental diagram is added to the figure for comparison purposes. The similarity between the numerical and experimental Bode diagrams demonstrates the representativeness of the obtained FE model.

5.1 Frequency-domain analysis

In the present analysis, the uncertain envelope of the FRF was obtained by considering Young's modulus of the shaft as uncertain information. Regarding the stochastic approach, uncertain Young's modulus is modeled as a Gaussian random field with nominal value E_s = 205 GPa and a 15% dispersion level. The convergence analysis was carried out to evaluate the number of terms retained in the truncated KL expansion (n_{KL}) and the number of samples for MCS (n_S). The RMS convergence analysis for the realizations of the FRF is assessed according to Eq. (6). **Figure 10** presents the obtained results. Note that convergence was achieved for n_{KL} = 40 and n_S = 250.

For the fuzzy approach, a fuzzy triangular number with the same nominal value and dispersion considered for the stochastic approach ($\tilde{E}_S$ = 205 $\pm$ 15% GPa) is used. The objective function of the α-level optimization is the norm of the FRF.

In this contribution, the performed uncertainty analysis aims at obtaining the minimum and maximum responses of the rotor system, i.e., the bounds of the uncertain dynamic responses. Therefore, the fuzzy uncertainty analysis was devoted to the α-level, α_k = 0. Thus, the dynamic responses of the rotor are obtained

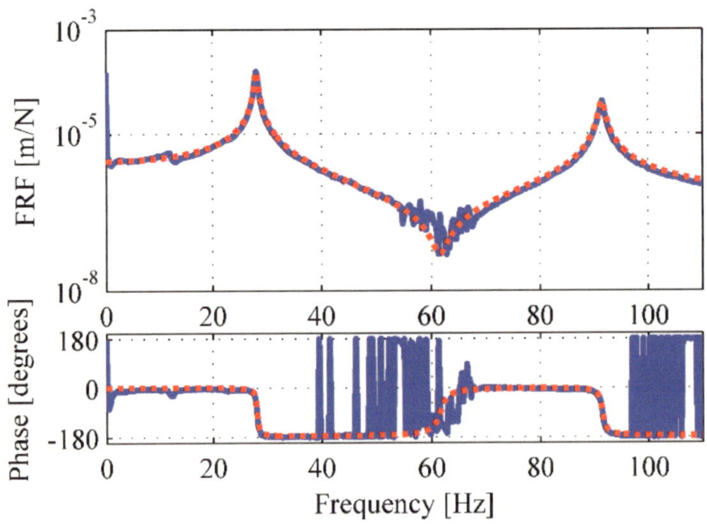

Figure 9.
Simulated (▬ ▬ ▬) and experimental (▬ ▬ ▬) Bode diagrams obtained from impact forces applied at D_1 and the sensor S_{8X} [13].

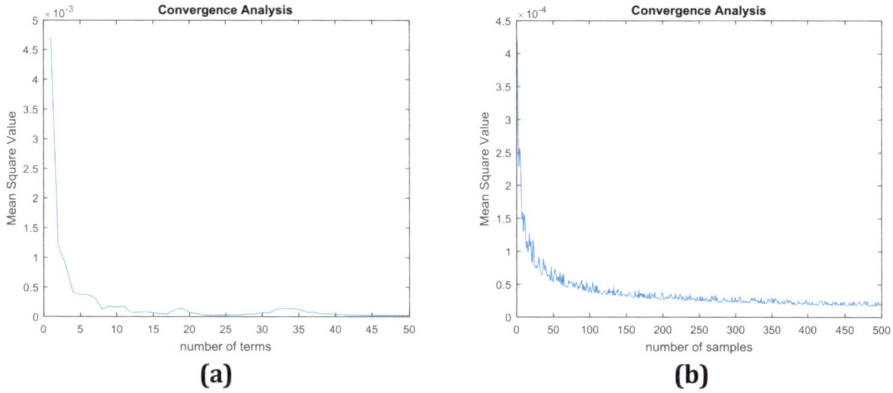

Figure 10.
Convergence analysis for the FRF: (a) n_{KL} and (b) n_s [13].

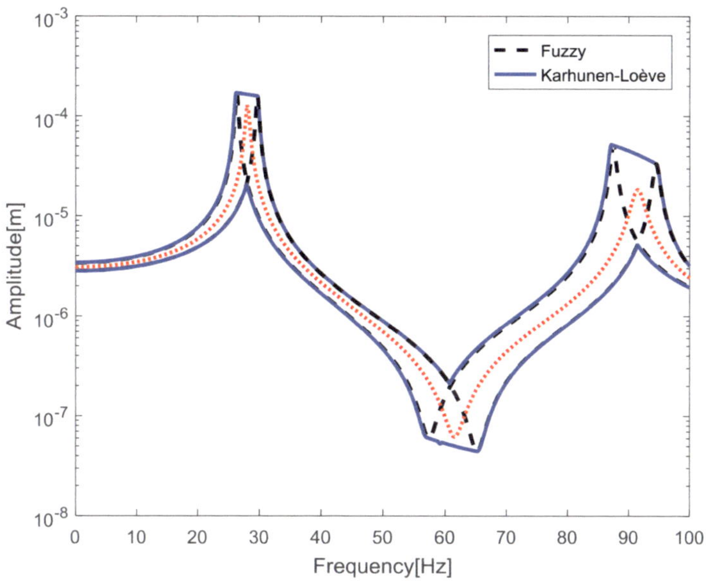

Figure 11.
FRFs obtained by using the stochastic and fuzzy approaches [13].

by considering the maximum level of uncertainty. Moreover, the stochastic approach is also applied to compute the minimum and maximum dynamic responses of the rotor.

Figure 11 presents a comparative evaluation of the FRFs' uncertain envelopes obtained by applying the stochastic and fuzzy approaches. In this case, the obtained FRFs were determined by considering the force applied along the x direction of disc D_1 and sensor S_{8X}. The results show that the uncertain envelopes obtained from the stochastic and fuzzy approaches are similar. Additionally, the updated FRF is also shown for comparison purposes.

5.2 Time-domain analysis

The time-domain analysis was performed based on the orbits of the flexible shaft. This analysis considers uncertainties affecting the stiffness coefficients k_{xx}

and k_{zz} of bearing B_1. For the stochastic approach, the uncertain parameters were modeled as Gaussian random variables with $k_{xx} = 8.551 \times 10^5$ N/m, $k_{zz} = 1.198 \times 10^6$ N/m (mean values), and deviation of ±10%. The rotation speed of the rotor is 1200 rev/min, and an unbalance of 487.5 g mm/0^0 was applied to disc D_1.

The convergence analysis was performed to determine n_{KL} and n_s based on the time-domain vibration responses of the rotor system. **Figure 12** shows the obtained results. Note that convergence was achieved for $n_{KL} \geq 100$ and $n_s \geq 500$.

Considering the fuzzy approach, the uncertain parameter is defined as a fuzzy triangular number with the same nominal value and deviation of the stochastic modeling. The objective function of the α-level optimization is written as the norm of the shaft displacement measured by sensor S_{8X}. **Figure 13** presents a comparative

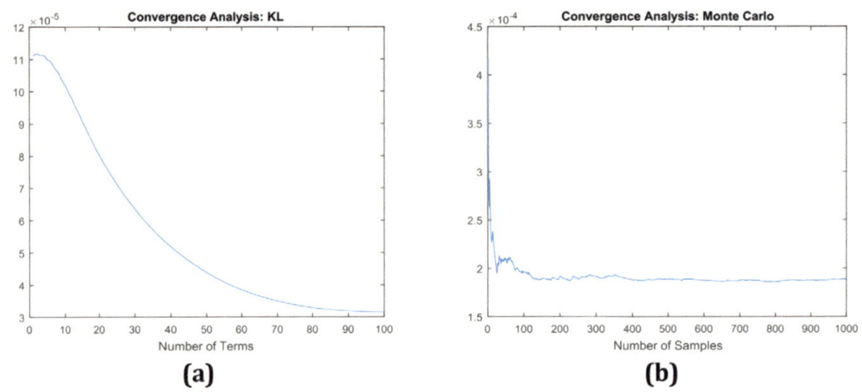

Figure 12.
Convergence analysis for the orbits: (a) n_{KL} and (b) n_s [13].

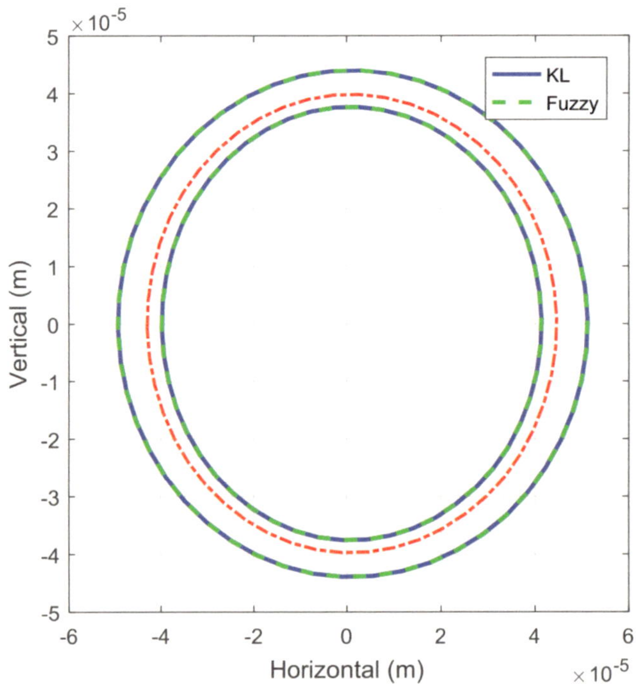

Figure 13.
Orbits obtained by using both the stochastic and fuzzy approaches [13].

evaluation of the uncertain envelopes of the rotor orbits determined by using the stochastic and fuzzy approaches. Note that the obtained results are similar, demonstrating that both approaches lead to equivalent responses.

6. Conclusions

This chapter is dedicated to the modeling, numerical methods, and simulations for the uncertainty analysis of flexible rotors. The stochastic and fuzzy approaches showed to be suitable methods to quantify the effect of uncertain parameters on the dynamic responses of rotating machines. The comparative study permitted to evaluate the two studied approaches is based on numerical simulations. Although the numerical results obtained by applying both approaches were similar, the fuzzy approach demands a greater computational effort than the stochastic method. Nevertheless, the stochastic approach requires an extensive mathematical background and an insight knowledge on the uncertain parameters. In this case, the stochastic distribution should be known or assumed. However, both approaches can be applied to the design of rotating machines.

Acknowledgements

The authors are thankful for the financial support provided to the present research effort by CNPq (574001/2008-5, 304546/2018-8, and 431337/2018-7), FAPEMIG (TEC-APQ-3076-09, TEC-APQ-02284-15, TEC-APQ-00464-16, and PPM-00187-18), and CAPES through the INCT-EIE. The authors are also thankful to the companies CERAN, BAESA, ENERCAN, and Foz do Chapecó for the financial support through the R&D project Robust Modeling for the Diagnosis of Defects in Generating Units (02476-3108/2016).

Author details

Fabian Andres Lara-Molina[1], Arinan De Piemonte Dourado[2], Aldemir Ap. Cavalini Jr[3] and Valder Steffen Jr[3]*

1 Mechanical Engineering Department, Federal University of Technology-Paraná, Cornélio Procópio, Brazil

2 Department of Mechanical and Aerospace, Engineering Probabilistic Mechanics Laboratory, University of Central Florida, Orlando, FL, USA

3 LMEst – Structural Mechanics Laboratory, Federal University of Uberlândia, School of Mechanical Engineering, Brazil

*Address all correspondence to: vsteffen@ufu.br

IntechOpen

© 2019 The Author(s). Licensee IntechOpen. This chapter is distributed under the terms of the Creative Commons Attribution License (http://creativecommons.org/licenses/by/3.0), which permits unrestricted use, distribution, and reproduction in any medium, provided the original work is properly cited.

References

[1] Didier J, Faverjon B, Sinou JJ. Analyzing the dynamic response of a rotor system under uncertain parameters by polynomial chaos expansion. Journal of Vibration and Control. 2012;**18**(5):587-607

[2] Koroishi EH, Cavalini AA Jr, de Lima AM, Steffen V Jr. Stochastic modeling of flexible rotors. Journal of the Brazilian Society of Mechanical Sciences and Engineering. 2012;**34** (SPE2):574-583

[3] Lara-Molina FA, Koroishi EH, Steffen V Jr. Uncertainty analysis of flexible rotors considering fuzzy parameters and fuzzy-random parameters. Latin American Journal of Solids and Structures. 2015;**12**(10): 1807-1823

[4] Cavalini AA, Silva ADG, Lara-Molina FA, Steffen V. Dynamic analysis of a flexible rotor supported by hydrodynamic bearings with uncertain parameters. Meccanica. 2017;**52**(11–12):2931-2943

[5] Lalanne M, Ferraris G. Rotordynamics—Prediction in Engineering. New York: John Wiley & Sons; 1998

[6] Maia NMM, Montalvão e Silva JM. Theoretical and Experimental Modal Analysis. England: Research Studies Press LTD.; 1997. 468 p

[7] Soize C. Stochastic modeling of uncertainties in computational structural dynamics—Recent theoretical advances. Journal of Sound and Vibration. 2013;**332**(10):2379-2395

[8] Ghanem RG, Spanos PD. Stochastic finite element method: Response statistics. In: Stochastic Finite Elements: A Spectral Approach. New York, NY: Springer; 1991. pp. 101-119

[9] Florian A. An efficient sampling scheme: updated Latin hypercube sampling. Probabilistic Engineering Mechanics. 1992;**7**(2):123-130

[10] Möller B, Beer M. Fuzzy Randomness: Uncertainty in Civil Engineering and Computational Mechanics. Springer Science & Business Media; 2013

[11] Cavalini AA, Silva AG, Lara-Molina FA, Steffen V. Uncertainty analysis of a tilting-pad journal bearing using fuzzy logic techniques. Journal of Vibration and Acoustics. 2016;**138**(6):061016

[12] Price KV, Storn RM, Lampinen JA. Differential Evolution: A Practical Approach to Global Optimization. Berlin, Germany: Springer-Verlag; 2005

[13] Silva ADG, Cavalini AA, Steffen V. Uncertainty quantification techniques applied to rotating systems: A comparative study. Journal of Vibration and Control. 2017;**1**(1):1-16

Chapter 3

Advance Measurement Techniques in Turbomachines

Fangyuan Lou

Abstract

This chapter focuses on advanced measurement techniques that have been used in applications of turbomachines including temperature measurements, pressure measurements, velocity measurements, and strain/stress measurements. Though the measurement techniques are fundamentally the same as those used in other applications, the unique features associated with turbomachines place challenges in implementing these techniques. This chapter covers the fundamental working principles of individual measurement technique as well as the highlights of its application in turbomachines.

Keywords: measurement techniques, temperature-sensitive paint, pressure-sensitive paint, laser Doppler velocimetry, particle image velocimetry, hot-wire anemometry, strain gauges, nonintrusive stress measurement systems

1. Introduction

Turbomachine consists a wide and diverse class of devices that have been used in air, land, and sea. Below is a list of representative applications for turbomachines:

- Fans and blowers
- Compressors in aviation (gas turbine engines), transportation (turbocharging systems), and oil and gas applications including axial, radial, mixed, and scroll compressors
- Turbines in gas turbine engines, steam turbines, and hydraulic turbines
- Wind turbines
- Pumps
- Propellers and open rotors

The flow in turbomachines is highly three-dimensional, turbulent, and inherently unsteady. The unsteady nature of the flow in turbomachines is a result of work exchange between the machine and its working fluid. These complex flow phenomena affect the performance and operability. The interactions between the flow and hardware structures can result in undesired noise, vibration, and sometimes failure of the machine. On one hand, enhanced understanding of the complex flow phenomena in turbomachines is essential for the development of better turbomachines in the future and, thus, requires experimental benchmark data associated with the flow including velocity, pressure, and temperature. On the other hand, better monitoring of the health status of the rotating groups (i.e., stress of rotors) is also of great importance for failure prevention.

Extensive experimental studies have been performed during the past few decades to investigate the complex flow as well as fluid-structure interactions in turbomachines. This chapter discusses the advanced techniques as well as highlights the challenges in implementing these techniques.

2. Velocity and turbulence measurements

This section discusses three techniques for velocity and turbulence measurements in applications of turbomachines including hot-wire anemometry, laser Doppler velocimetry (LDV), and particle image velocimetry (PIV). A brief introduction of the working principles, features associated with the measurement technique, and the challenges for implementation in turbomachines are presented. Lists of previous studies in the open literature applying these measurement techniques to turbomachines are also provided.

2.1 Thermal anemometer

Hot-wire anemometry is an intrusive measurement technique that provides instantaneous velocity and turbulence measurements. It allows characterizing high-frequency flow structures at relatively inexpensive cost when compared with alternative approaches such as laser Doppler velocimetry and other optical techniques. Below highlights the features of the hot-wire anemometry.

Intrusive velocity measurement. In contrast to techniques for the measurement of flow velocities employing probes such as pressure tubes or hot wires, the LDV technique features a nonintrusive nature eliminating the disturbances introduced by the presence of the probes.

Direct velocity measurement. The hot-wire anemometry does not require particle tracers and provides direct measurements of the fluid velocity and turbulence.

Point measurement. Hot-wire anemometry is a point-based measurement, and the measurement volume is determined by the dimensions of the employed wires.

A hot-wire probe consists of a short-length (on the magnitude of millimeter), fine-diameter (on the magnitude of micrometer) wire that is attached to two prongs. The technique relies on the changes in heat transfer between the heated wire and the fluid the wire is exposed. Heat is introduced in the sensor by Joule heating and is lost primarily by forced convection. A significant parameter that controls the operation of the sensor is the relative difference in temperature between the heated up wire and the flow, which is related to the overheat ratio of the sensor. Changes in the flow properties, such as velocity, density, temperature, and humidity, will cause a corresponding change in the heat transfer from the wire which can be detected and measured. Hot wires are typically run in either a constant temperature (CT) mode or constant current (CC) mode. A sketch of circuit for thermal anemometry operating in constant temperature mode is shown in **Figure 1**. Constant temperature anemometry utilizes a rapid-response servo circuit coupled with the Wheatstone bridge amplifier to control the applied voltage and maintain a constant wire resistance, which in turn maintains a constant wire temperature. It eliminates the effect of thermal inertial of the wire, as well as the system response time, and, thus, provides a better frequency response compared to the constant current

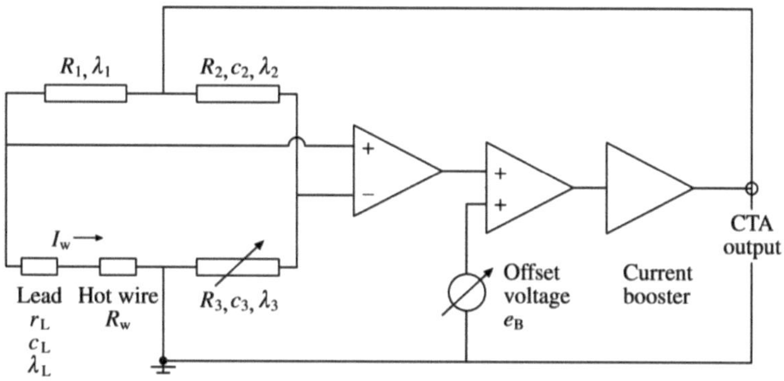

Figure 1.
Circuit diagram of a thermal anemometer operating at constant temperature mode.

operation. Most velocity and turbulence measurements are acquired in this manner. Attributed to the extremely small thermal mass of the wire, this technique allows detection of very-high-frequency fluctuations in the flow. This is another advantage offered by the hot-wire anemometry.

The hot wires can be configured in the manner of single wire, cross wires, and triple wires. The single wire is primarily useful for mean flow quantity measurements and is not as accurate as triple wire simultaneous measurements for turbulence and Reynolds stress measurements.

Because of the small size and high-frequency response offered by hot-wire anemometer, the technique has been used extensively in investigations of the flow fields in turbomachines. While early investigations using hot-wire/hot-film probes were mostly qualitative, substantial amounts of quantitative investigations have also been carried out in the past few decades on various aspects, using many different types of hot-wire and hot-film probes. **Table 1** provides a list of representative studies in applications of turbomachines.

2.2 Laser Doppler velocimetry (LDV)

Laser Doppler velocimetry is an optical nonintrusive technique which measures the instantaneous velocity at a given point in a flow field. It was first developed by Yeh and Cummins in 1964 and is now a well-established technique. This technique has been widely used in all kinds of fluid flow applications, including laminar flow, turbulent flow, flow inside turbomachinery, and flow inside combustion chambers. Because of its high accuracy, it is also used as a benchmark validation tool for planar velocimetry techniques (i.e., PIV, PTV). Below lists the features in the LDV measurement technique:

Nonintrusive velocity measurement. In contrast to techniques for the measurement of flow velocities employing probes such as pressure tubes or hot wires, the LDV technique features a nonintrusive nature eliminating the disturbances introduced by the presence of the probes.

Indirect velocity measurement. The LDV technique measures the velocity of a fluid element indirectly in most of cases by means of the measurement of the velocity of tracer particle being added to the flow.

Point measurement. The same as hot wire, the LDV is a point-based measurement technique that characterizes the velocity in its measurement volume.

A sketch of the LDV working principle is shown in **Figure 2**. The system shown in the sketch is a 1-D LDV system operated in backscattering mode, but the working theory is the same for all the LDV systems. A single laser beam is emitted from a

Author(s)	Year	Sensor type	Type of machine	Subject of study
Lakshminarayana and Poncet [1]	1974	Singe and cross wire	Axial inducer	Rotor wakes
Gorton and Lakshminarayana [2]	1976	Triple wire	Axial inducer	Mean flow and turbulence
Hah and Lakshminarayana [3]	1980	Triple wire	Axial compressor	Freestream turbulence on a rotor wake
Hodson et al. [4]	1994	Hot film	Low-pressure turbine	Rotation and blade incidence on rotor wake
Camp and Shin [5]	1995	Hot wire, hot film, single sensor	Axial compressor	Turbulence intensity and length scale
Witkowski et al. [6]	1996	Hot film (triple wire)	Axial compressor	3-D wake decay and secondary flows
Halstead et al. [7]	1997	Surface-mounted hot film/hot wire	Axial compressor and turbine	Unsteady boundary layer
Hsu and Wo [8]	1997	Slanted hot wire	Axial compressor	Unsteady wake
Ristic and Lakshminarayana [9]	1998	Cross wire	Axial turbine	3-D boundary layer
Furukawa et al. [10]	1998	Hot wire	Diagonal flow rotor	Tip flow field
Sentker and Riess [11]	2000	Split hot film	Axial compressor	Wake-blade interaction
Velarde-Suarez et al. [12]	2001	Cross wire	Centrifugal fan	Unsteady flow
Pinarbasi [13]	2008	Triple wire	Centrifugal compressor	Diffuser flow
Goodhand and Miller [14]	2011	Single wire	Axial compressor	Boundary layer development during spike-type stalling
Weichert and Day [15]	2014	Single wire	Axial compressor	Tip region flow during spike-type stalling

Table 1.
Representative studies that have used thermal anemometer technique in studying the flow field of turbomachines.

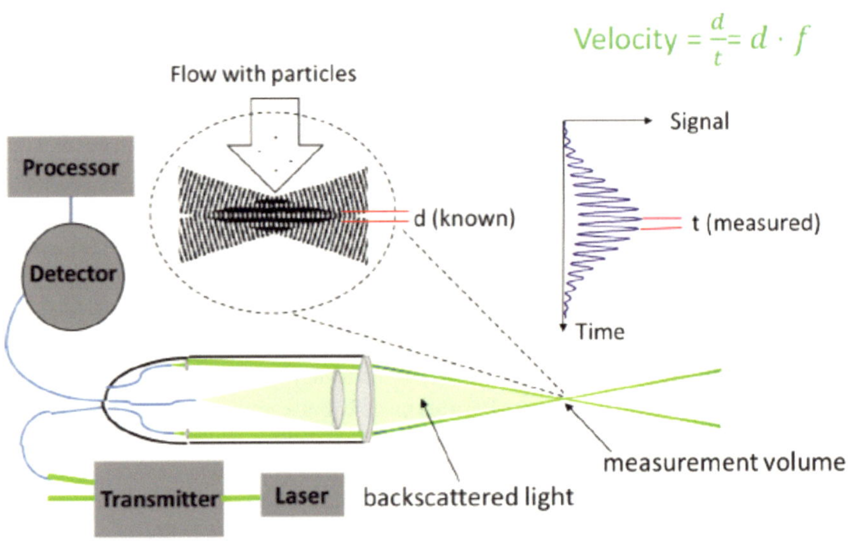

Figure 2.
Sketch of 1-D backscattering LDV system.

Author(s)	Year	Type	Type of machine	Subject of study
Wisler and Mossey [16]	1973	1-D	Axial compressor rotor	Rotor passage relative flow
Pierzga and Wood [17]	1985	1-D	Axial fan rotor	3-D flow field in a transonic rotor
Strazisar [18]	1985	1-D	Axial fan rotor	Flow structure in transonic fan rotor
Murthy and Lakshminarayana [19]	1986	1-D	Axial compressor	Rotor tip region flow
Beaudoin et al. [20]	1992	2-D	Centrifugal pump	Effects of orbiting impeller
Hathaway et al. [21]	1993	3-D	Centrifugal compressor	3-D flow structure
Farrell and Billet [22]	1994	na	Axial pump	Tip vortex cavitation
Fagan and Fleeter [23]	1994	1-D	Centrifugal compressor	Flow structure
Abramian and Howard [24]	1994	1-D	Centrifugal impeller	Impeller relative flow field
Zaccaria and Lakshminarayana [25]	1997	2-D	Axial turbine	Rotor passage flow field
Adler and Benyamin [26]	1999	2-D	Axial turbine	Stator wake propagation
Ristic et al. [27]	1999	3-D	Axial turbine	3-D flow field downstream of rotor turbine
Faure et al. [28]	2001	2-D	Axial compressor	Flow structure
Van Zante et al. [29]	2002	2-D	Axial compressor	Blade row interactions
Ibaraki et al. [30]	2003	2-D	Centrifugal impeller	Impeller flow
Higashimori et al. [31]	2004	2-D	Centrifugal impeller	Impeller flow field
Faure et al. [32]	2004	3-D	Axial compressor	Reynolds stresses measurements
Schleer et al. [33]	2004	2-D	Centrifugal compressor	Impeller discharge flow
Ibaraki et al. [34]	2009	2-D	Centrifugal impeller	Impeller flow field
Gooding et al. [35]	2019	3-D	Centrifugal compressor	Diffuser flow

Table 2.
Representative studies that have used LDV technique in studying the flow field of turbomachines.

laser head operating in continuous mode and then enters into the optical transmitter. Inside the transmitter, this single beam is split and frequency shifted using the beam splitter (BS), an achromatic lens, and a Bragg cell. Pairs of monochrome laser beams (depending on the number of velocity components needs to be measured: one pair for a 1-D system, two pairs for a 2-D system, and three pairs for a 3-D system) generated by the transmitter are then conveyed to the optical probes using fiber cables. The laser beams coming out of the probe intersect at the focal point of the front lens. At this focal point, at which the measurement volume is located, an ellipsoidal volume with bright and dark fringe patterns is formed by the interference of the laser beams. As flow particles traverse through this measurement volume, the backscattered light is collected by the receiving optics inside the probe and further processed by a burst spectrum analyzer. Inside the spectrum analyzer, the time intervals for the burst traveling through the bright and dark patterns are measured. Those measured time intervals, combined with the known distance between the adjacent bright and dark strips, yield the calculation of velocity.

This nonintrusive feature of the technique attracted the attention of experimentalist in the field of turbomachines soon after it was introduced in the 1960s. In addition to being nonintrusive, it allowed the velocity measurements in the rotating reference frame without having to use complex rotating probe traverse or data transmission mechanisms (i.e., slip ring or telemetry system). A list of representative studies using LDV for measurements in turbomachines is provided in **Table 2**.

One of the earliest applications of LDV to turbomachinery was conducted by Wisler and Mossey [16] to measure the relative velocity across the first-stage rotor blade row using a single-component LDV system. The flow was seeded by spray atomizing a dilute water suspension of 1-μm-diameter polystyrene latex particles. A sketch of the experimental setup and a sample contour plot of relative velocity within the rotor passage at mid-span (50%) are shown in **Figure 3**. In addition, a sketch of experimental setup for three-component LDV in a centrifugal compressor is presented in **Figure 4**. As summarized in the table, majority of the investigations involving LDV have been performed in a stationary frame of reference. To measure the flow field in rotors, the rotor passage period has been discretized into bins, each with a finite time interval. The results in each bin were ensemble averaged to obtain the mean velocity and turbulence parameters across the rotor passage. To reach convergence in mean velocity and turbulence parameters, a large data set per bin is favored which requires a larger bin size. However, an increase in the bin size introduces the effects of spatial variations in the flow structure. An alternative approach is to conduct measurements in the rotating frame of reference. However, this makes the experimental very challenging, and the only study reported in the open literature of this category was performed by Abramian and Howard [24]. The experiment was conducted in a centrifugal impeller using a Dove prism to transfer the laser beams to the rotating frames of reference.

There are challenges in implementing LDV to turbomachines. Generally speaking, the challenges can be categorized into the optical accessibility-related issues and particle related. Typically, the three-dimensional twisted rotor blades make it difficult to shine laser beams to the interested measurement locations and require the LDV system operating in backscatter mode. Comparing to the favorable forward-scatter configuration, the signal-to-noise ratio of backscatter mode is commonly one to three orders of magnitude smaller. Additionally, the signal-to-noise ratio gets further deteriorated at measurement locations close to metal surfaces due to reflections and in applications of curved optical windows due to the distortion of laser beams through the windows. These distortions increase the uncertainty of the measurements by deforming the measurement volume and changing the measurement location. It is also challenging to deliver particles to target measurement locations due to the strong secondary flow in turbomachines.

2.3 Particle image velocimetry (PIV)

In addition to LDV, particle image velocimetry is another *nonintrusive* technique for velocity measurements. The same as LDV, PIV is also an *indirect* measurement technique and requires tracer particles. Different from LDV and thermal anemometer which are point-based measurement technique, PIV offers *full-field* measurements and allows mapping of large parts of flow field. The working principle of PIV is schematically described in **Figure 5**. The principle of PIV is based on the measurement of the displacement of small tracer particles during a short time interval. This indirect measurement nature requires the tracer particles to be sufficiently small to precisely follow the motion of fluid. The tracer particles are typically illuminated using a thin light sheet generated from pulsed laser head. A pair of images for the illuminated flow field is taken by a digital imaging device, typically a CCD camera. Depending on the number and configuration of camera employed, either 2-D or 3-D flow field could be obtained using cross-correlation analysis to measure the displacement

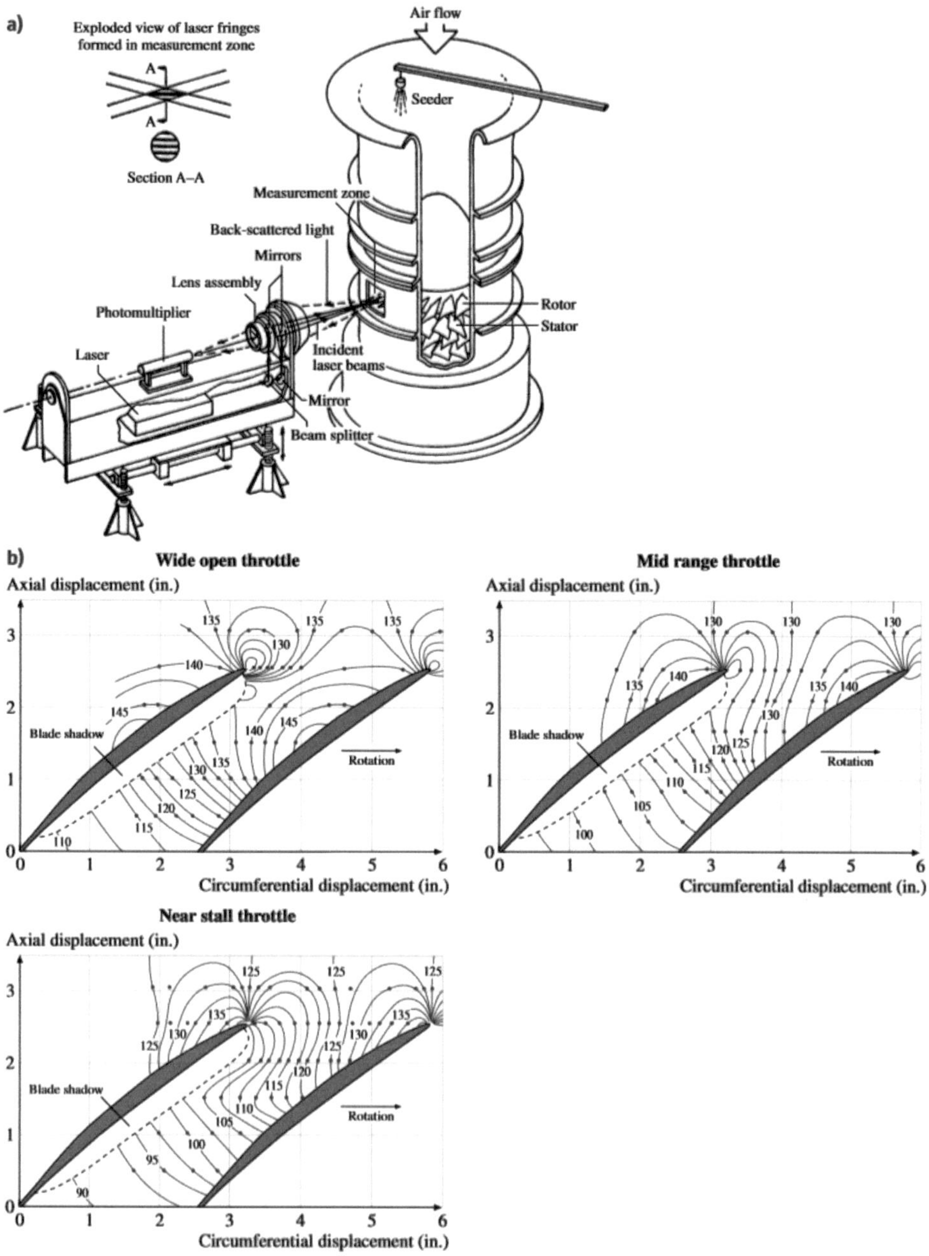

Figure 3.
The LDV setup (a) and measured relative velocity contours within the rotor passage of a low-speed research compressor (b) Wisler and Mossey [16].

of particles in each small interrogation areas. A single-camera system allows characterization of the two velocity components within the measurement plane, while stereo imaging using two inclined cameras provides all three components of the velocity in the illuminated plane.

Effort of implementing PIV in investigations of turbomachinery flow filed has been entertained since the emergence of the technique. Previous researchers have performed both two-dimensional and stereoscopic PIV measurements within various axial and centrifugal turbomachinery facilities. A few highlights of selected previous research are presented, and a more extended set of references is provided in **Table 3**.

Figure 4.
Photo of experimental setup for study of flow in a centrifugal compressor using three-component LDV.

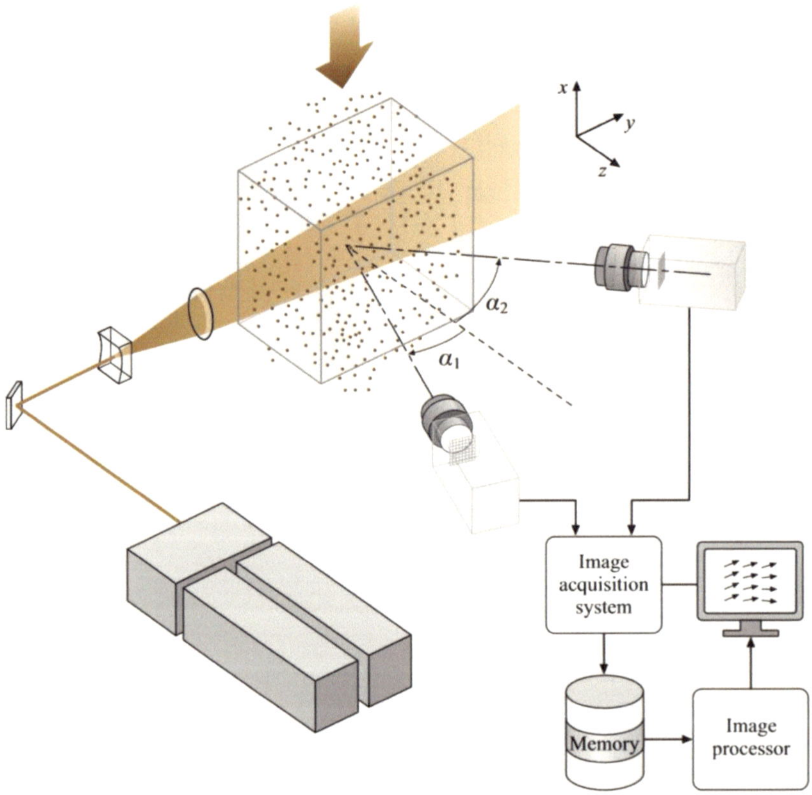

Figure 5.
Sketch of typical PIV setup [36].

Figure 6 presents sample results from the two-dimensional measurements performed in an axial pump at Johns Hopkins University. The distribution of phase-averaged velocity, vorticity, and turbulent kinetic energy at the mid-span of the second stage was characterized [37]. **Figure 7** shows sample data obtained in a high-speed centrifugal compressor operating both at the design point and during surge [43, 46]. As summarized in the table, majority of these studies insert a periscopic optical probe into the flow for light sheet delivery. This results in invasive measurement and also

DOI: http://dx.doi.org/10.5772/intechopen.85910

Author(s)	Year	Type	Type of machine	Subject of study
Paone et al. [38]	1989	2-D	Centrifugal pump	Flow structure
Chu et al. [39, 40]	1995	2-D	Centrifugal pump	Unsteady flow and pressure fluctuations
Day et al. [41]	1996	2-D	Axial turbine	Effect of film cooling on flow structure
Dong et al. [42]	1997	2-D	Centrifugal pump	Unsteady flow and noise
Wernet [43]	2000	2-D	Centrifugal compressor	Diffuser flow structure
Sinha and Katz [44]	2000	2-D	Centrifugal pump	Diffuser flow field
Uzol and Camci [45]	2001	2-D	Axial turbine cascade	Trailing edge coolant ejection
Wernet et al. [46]	2001	2-D	Centrifugal compressor	Diffuser flow during surge
Chow et al. [37]	2002	2-D	Axial pump	Wake-wake interactions
Uzol et al. [47]	2002	2-D	Axial pump	Unsteady flow and deterministic stresses
Sanders et al. [48]	2002	2-D	Axial compressor	Blade row interactions
Estevadeordal et al. [49]	2002	2-D	Axial compressor	Wake-blade interactions
Woisetschlager et al. [50]	2003	2-D	Axial turbine cascade	Turbine wake
Uzol et al. [51]	2003	3-D	Axial pump	3-D wake structure and tip vortex
Lee et al. [52]	2004	3-D	Marine propeller	Propeller wake
Wernet et al. [53]	2005	3-D	Axial compressor	Tip region flow
Yu and Liu [54]	2006	3-D	Axial compressor	Unsteady flow
Ibaraki et al. [55]	2007	2-D	Centrifugal compressor	Unsteady diffuser flow
Estevadeordal et al. [56]	2007	3-D	Axial compressor	Wake-rotor interactions
Voges et al. [57]	2007	2-D	Centrifugal compressor	Diffuser flow
Voges et al. [58]	2012	3-D	Axial compressor	Tip region flow
Guillou et al. [59]	2012	3-D	Turbocharger compressor	Impeller inlet flow
Gancedo et al. [60]	2016	3-D	Turbocharger compressor	Impeller inlet flow
Bhattacharya et al. [61]	2016	3-D	Axial compressor	Rotor flow field

Table 3.
Representative studies that have used PIV technique in studying the flow field of turbomachines.

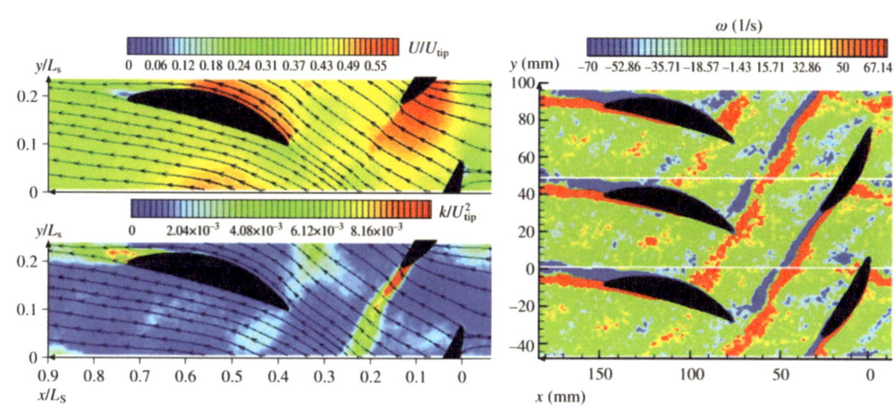

Figure 6.
Sample PIV data obtained in an axial pump facility at Johns Hopkins University: Phase-averaged velocity field (top left), turbulent kinetic energy (bottom left), and vorticity (right) at mid-span within an entire stage [37].

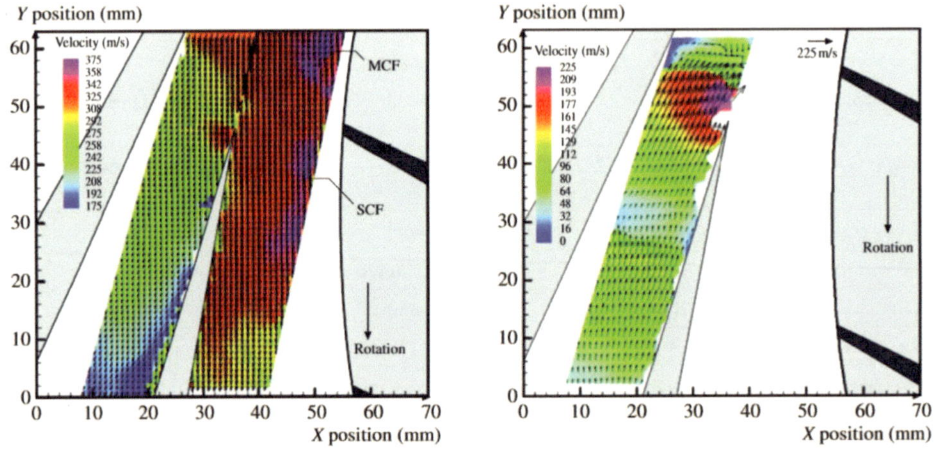

Figure 7.
Sample PIV measurements in the diffuser passage of a high-speed centrifugal compressor at both the design point (left) [43] and during a surge (right) [46].

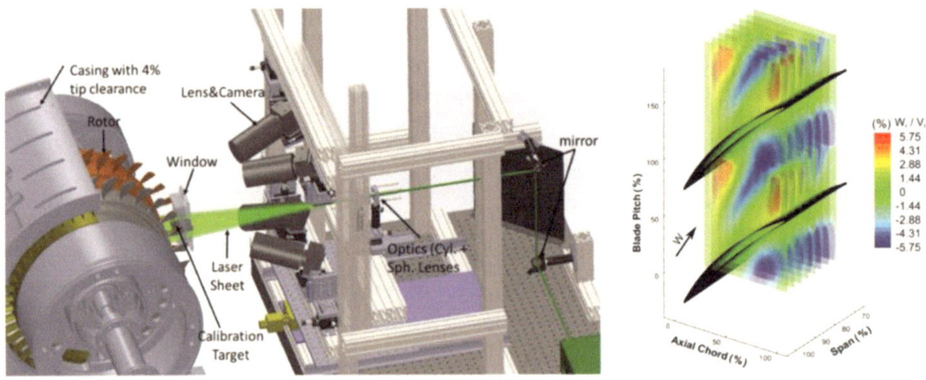

Figure 8.
Experimental setup for PIV measurements performed at Purdue University (left) and sample results (right) of normalized radial velocity at fixed spanwise locations for stereo reconstructed velocity field [61].

significantly limits the region of flow field that can be imaged. To address these challenges, a new approach was introduced in a recent study performed at Purdue University in a multistage axial compressor [61]. The same window was used for both laser sheet delivery and image recording. By doing so, it eliminates the presence of invasive probe for light sheet delivery. A sketch of the experimental setup and sample results is shown in **Figure 8**. As illustrated in the figure, the PIV measurements were performed in the second-stage rotor passage (rotor 2). To eliminate light reflections from the blade surface and hub, fluorescent dye with sufficiently separated absorption and emission wavelengths was introduced with the seeding fluid, and lens filters blocking wavelengths below 540 nm were used to filter laser reflections. Slices of normalized radial velocity at fixed spanwise positions were presented to illustrate the development of the tip leakage flow across the rotor passage.

3. Pressure-sensitive paints

Conventionally, surface pressures are measured using hundreds of pressure taps or flush-mounted transducers to obtain a reasonable spatial distribution. This makes

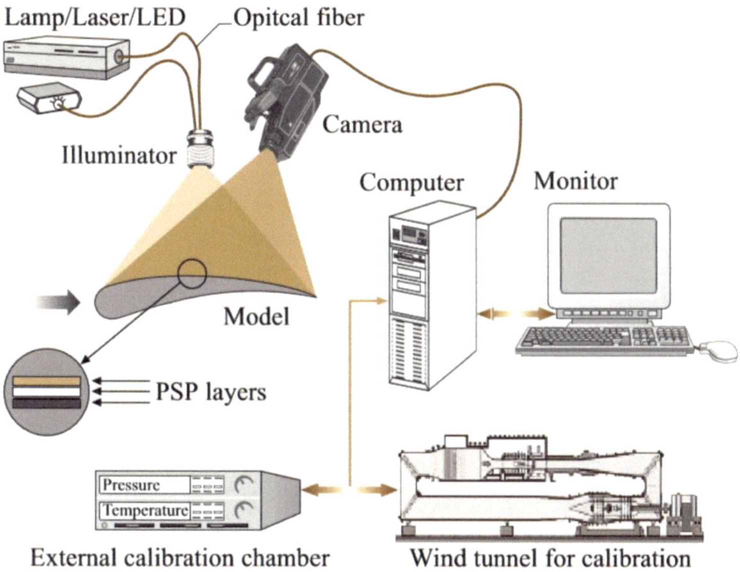

Figure 9.
Schematic of pressure-sensitive paint measurement system [36].

Author(s)	Year	Type of machine	Subject of study
Sabroske et al. [63]	1995	Axial compressor	Blade pressure distribution
Liu et al. [64]	1997	High-speed axial compressor	Blade surface pressure
Navarra [65]	1997	Axial compressor	Blade surface pressure
Bencic [66]	1998	Axial fan	Blade surface pressure
Engler et al. [67]	2000	Axial turbine	Shock movement and corner stall
Navarra et al. [68]	2001	Axial compressor	Blade surface pressure in transonic conditions
Gregory et al. [69]	2002	Centrifugal compressor	Blade surface pressure
Lepicovsky and Bencic [70]	2002	Supersonic through flow fan	Effect of change operating conditions
Gregory [71]	2004	Centrifugal compressor	Effect of inlet distortion on surface pressures
Suryanarayanan et al. [72]	2010	Axial turbine	Filming cooling
Narzary et al. [73]	2012	Axial turbine	Effect of coolant density on turbine film cooling

Table 4.
Representative studies that have used PSP technique in turbomachines.

the measurements time-consuming and expensive. Recently, the introduction of pressure-sensitive paint (PSP) provides a new method for surface pressure measurement. Comparing to the conventional approaches by means of pressure taps or transducers which can only provide data at discrete points and are limited by installation locations, the PSP technique is very attractive; hence, it provides high-spatial-resolution pressure measurements without taps or transducers. The PSP technique is based on covering a surface with luminescent coatings. The luminescence of the coating is dependent on surface static pressure. With proper illumination, the surface pressure

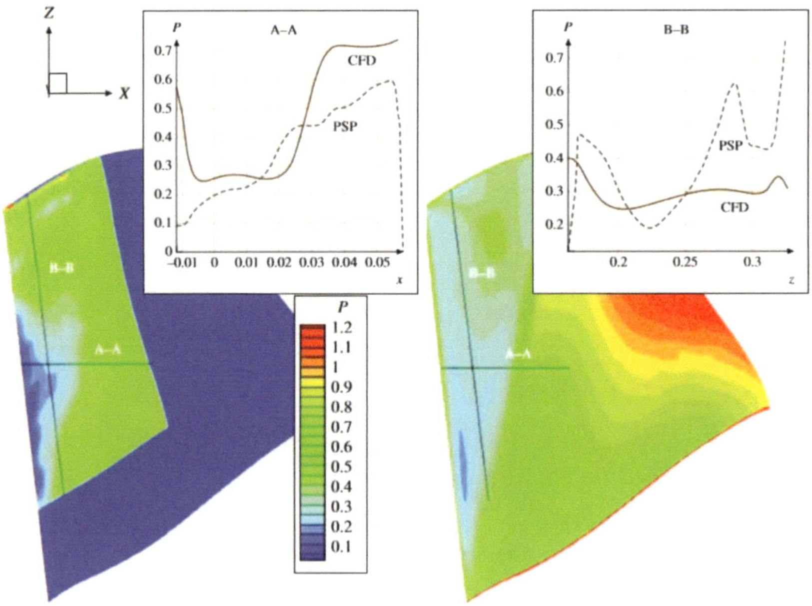

Figure 10.
Comparison of surface pressure distribution on a rotor suction side in a transonic axial compressor obtained using PSP (left) to CFD predictions (right) [68].

distribution is obtained from images of illuminated surface. **Figure 9** shows all the essential optical and electrical components of a PSP system. It consists of various illumination devices, a local image and data-acquisition system, and an external calibration chamber.

The first aerodynamic study using PSP is performed by Pervushin et al. in 1985 to measure the pressure of air on the surface of wind tunnel models [62], and since then, numerous studies using PSP in external aerodynamics research have been conducted. However, unlike the well-established applications in external aerodynamic research, the application of PSP in turbomachines is quite limited. **Table 4** lists the studies in the open literature that have used pressure-sensitive paint in turbomachines. A sample result of the PSP measurements together with the comparison to CFD results from the study conducted by Navarra et al. [68] is shown in **Figure 10**. The PSP measurement was conducted on the suction surface of the first-stage rotor of a state-of-the-art, full-scale transonic compressor.

4. Conclusions

This chapter attempts to provide a comprehensive but brief summary of several advanced measurement techniques that have been used in turbomachines. For each measurement technique, the fundamental working principle was provided first and followed by discussion of its application in turbomachines. A list of representative research from the open literature was also provided for reference.

Conflict of interest

The author claims there is no conflict of interest.

DOI: http://dx.doi.org/10.5772/intechopen.85910

Author details

Fangyuan Lou
Purdue University, West Lafayette, USA

*Address all correspondence to: louf@purdue.edu

IntechOpen

© 2020 The Author(s). Licensee IntechOpen. This chapter is distributed under the terms of the Creative Commons Attribution License (http://creativecommons.org/licenses/by/3.0), which permits unrestricted use, distribution, and reproduction in any medium, provided the original work is properly cited.

References

[1] Lakshminarayana B, Poncet A. A method of measuring three-dimensional rotating wales behind turbomachinery rotors. Journal of Fluids Engineering. 1974;**96**(2):87-91

[2] Gorton CA, Lakshminarayana B. A method of measuring the three-dimensional mean flow and turbulence quantities inside a rotating turbo-machinery passage. Journal of Engineering for Power. 1976;**98**(2):137-144

[3] Hah C, Lakshminarayana B. Freestream turbulence effects on the development of a rotor wake. AIAA Journal. 1981;**19**(6):724-730

[4] Hodson HP, Huntsman I, Steele AB. An investigation of boundary layer development in a multistage LP turbine. Journal of Turbomachinery. 1994;**116**(3):375-383

[5] Camp TR, Shin HW. Turbulence intensity and length scale measurements in multistage compressors. Journal of Turbomachinery. 1995;**117**(1):38-46

[6] Witkowski AS, Chmielniak TJ, Strozik MD. Experimental study of a 3D wake decay and secondary flows behind a rotor blade row of a low speed compressor stage. In: ASME 1996 International Gas Turbine and Aeroengine Congress and Exhibition; 10 June 1996; American Society of Mechanical Engineers. 1996. p. V001T01A107

[7] Halstead DE, Wisler DC, Okiishi TH, Walker GJ, Hodson HP, Shin HW. Boundary layer development in axial compressors and turbines: Part 1 of 4—Composite picture. In: ASME 1995 International Gas Turbine and Aeroengine Congress and Exposition; 5 June 1995; American Society of Mechanical Engineers. 1995. p. V001T01A109

[8] Hsu ST, Wo AM. Near-wake measurement in a rotor/stator axial compressor using slanted hot-wire technique. Experiments in Fluids. 1997;**23**(5):441-444

[9] Ristic D, Lakshminarayana B. Three-dimensional blade boundary layer and endwall flow development in the nozzle passage of a single stage turbine. Journal of Fluids Engineering. 1998;**120**(3):570-578

[10] Furukawa M, Saiki K, Nagayoshi K, Kuroumaru M, Inoue M. Effects of stream surface inclination on tip leakage flow fields in compressor rotors. In: ASME 1997 International Gas Turbine and Aeroengine Congress and Exhibition; 2 June 1997; American Society of Mechanical Engineers. 1997. p. V001T03A006

[11] Sentker A, Riess W. Experimental investigation of turbulent wake-blade interaction in axial compressors. International Journal of Heat and Fluid Flow. 2000;**21**(3):285-290

[12] Velarde-Suárez S, Ballesteros-Tajadura R, Santolaria-Morros C, González-Pérez J. Unsteady flow pattern characteristics downstream of a forward-curved blades centrifugal fan. Journal of Fluids Engineering. 2001;**123**(2):265-270

[13] Pinarbasi A. Experimental hot-wire measurements in a centrifugal compressor with vaned diffuser. International Journal of Heat and Fluid Flow. 2008;**29**(5):1512-1526

[14] Goodhand MN, Miller RJ. Compressor leading edge spikes: A new performance criterion. Journal of Turbomachinery. 2011;**133**(2):021006

[15] Weichert S, Day I. Detailed measurements of spike formation in an axial compressor. Journal of Turbomachinery. 2014;**136**(5):051006

[16] Wisler DC, Mossey PW. Gas velocity measurements within a compressor rotor passage using the laser Doppler velocimeter. Journal of Engineering for Power. 1973;**95**(2):91-96

[17] Pierzga MJ, Wood JR. Investigation of the three-dimensional flow field within a transonic fan rotor: Experiment and analysis. Journal of Engineering for Gas Turbines and Power. 1985;**107**(2):436-448

[18] Strazisar AJ. Investigation of flow phenomena in a transonic fan rotor using laser anemometry. Journal of Engineering for Gas Turbines and Power. 1985;**107**(2):427-435

[19] Murthy KN, Lakshminarayana B. Laser Doppler velocimeter measurement in the tip region of a compressor rotor. AIAA Journal. 1986;**24**(5):807-814

[20] Beaudoin RJ, Miner SM, Flack RD. Laser velocimeter measurements in a centrifugal pump with a synchronously orbiting impeller. In: ASME 1990 International Gas Turbine and Aeroengine Congress and Exposition; 11 June 1990; American Society of Mechanical Engineers. 1990. p. V001T01A087

[21] Hathaway MD, Chriss RM, Wood JR, Strazisar AJ. Experimental and computational investigation of the NASA low-speed centrifugal compressor flow field. In: ASME 1992 International Gas Turbine and Aeroengine Congress and Exposition; 1 June 1992; American Society of Mechanical Engineers. 1992. p. V001T01A074

[22] Farrell KJ, Billet ML. A correlation of leakage vortex cavitation in axial-flow pumps. Journal of Fluids Engineering. 1994;**116**(3):551-557

[23] Fagan JR, Fleeter S. Comparison of optical measurement techniques for turbomachinery flow fields. Journal of Propulsion and Power. 1994;**10**(2):176-182

[24] Abramian M, Howard JH. A rotating laser-Doppler anemometry system for unsteady relative flow measurements in model centrifugal impellers. Journal of Turbomachinery. 1994;**116**(2):260-268

[25] Zaccaria MA, Lakshminarayana B. Unsteady flow field due to nozzle wake interaction with the rotor in an axial flow turbine: Part I—Rotor passage flow field. In: ASME 1995 International Gas Turbine and Aeroengine Congress and Exposition; 5 June 1995; American Society of Mechanical Engineers. 1995. p. V001T01A081

[26] Adler D, Benyamin R. Experimental investigation of the stator wake propagation inside the flow passages of an axial gas turbine rotor. International Journal of Turbo and Jet-Engines. 1999;**16**(4):193-206

[27] Ristic D, Lakshminarayana B, Chu S. Three-dimensional flow field downstream of an axial-flow turbine rotor. Journal of Propulsion and Power. 1999;**15**(2):334-344

[28] Faure TM, Michon GJ, Miton H, Vassilieff N. Laser Doppler anemometry measurements in an axial compressor stage. Journal of Propulsion and Power. 2001;**17**(3):481-491

[29] Van Zante DE, To WM, Chen JP. Blade row interaction effects on the performance of a moderately loaded NASA transonic compressor stage. In: ASME Turbo Expo 2002: Power for Land, Sea, and Air; 1 January 2002; American Society of Mechanical Engineers. 2002. p. 969, 980

[30] Ibaraki S, Matsuo T, Kuma H, Sumida K, Suita T. Aerodynamics of a transonic centrifugal compressor impeller. Journal of Turbomachinery. 2003;**125**(2):346-351

[31] Higashimori H, Hasagawa K, Sumida K, Suita T. Detailed flow study of Mach number 1.6 high transonic

flow with a shock wave in a pressure ratio 11 centrifugal compressor impeller. Journal of Turbomachinery. 2004;**126**(4):473-481

[32] Faure TM, Miton H, Vassilieff N. A laser Doppler anemometry technique for Reynolds stresses measurement. Experiments in Fluids. 2004;**37**(3):465-467

[33] Schleer M, Hong SS, Zangeneh M, Roduner C, Ribi B, Pløger F, et al. Investigation of an inversely designed centrifugal compressor stage—Part II: Experimental investigations. Journal of Turbomachinery. 2004;**126**(1):82-90

[34] Ibaraki S, Sumida K, Suita T. Design and off-design flow fields of a transonic centrifugal compressor impeller. In: ASME Turbo Expo 2009: Power for Land, Sea, and Air; 1 January 2009; American Society of Mechanical Engineers. 2009. pp. 1375-1384

[35] Gooding JW, Fabian JC, Key NL. LDV characterization of unsteady vaned diffuser flow in a centrifugal compressor. In: ASME Turbo Expo 2019: Turbomachinery Technical Conference & Exposition; 17 June 2019; American Society of Mechanical Engineers. 2019. p. GT2019-90476

[36] Tropea C, Yarin AL. Springer Handbook of Experimental Fluid Mechanics. Berlin, Heidelberg: Springer Science & Business Media; 2007

[37] Chow YC, Uzol O, Katz J. Flow nonuniformities and turbulent "hot spots" due to wake-blade and wake-wake interactions in a multi-stage turbomachine. Journal of Turbomachinery. 2002;**124**(4):553-563

[38] Paone N, Riethmuller ML, Van den Braembussche RA. Experimental investigation of the flow in the vaneless diffuser of a centrifugal pump by particle image displacement velocimetry. Experiments in Fluids. 1989;**7**(6):371-378

[39] Chu S, Dong R, Katz J. Relationship between unsteady flow, pressure fluctuations, and noise in a centrifugal pump—Part A: Use of PDV data to compute the pressure field. Journal of Fluids Engineering. 1995;**117**:24-29

[40] Chu S, Dong R, Katz J. Relationship between unsteady flow, pressure fluctuations, and noise in a centrifugal pump—Part B: Effects of blade-tongue interactions. Journal of Fluids Engineering. 1995;**117**:30-35

[41] Day K, Lawless P, Fleeter S. Particle image velocimetry measurements in a low speed two stage research turbine. In: 32nd Joint Propulsion Conference and Exhibit. 1996. p. 2569

[42] Dong R, Chu S, Katz J. Effect of modification to tongue and impeller geometry on unsteady flow, pressure fluctuations, and noise in a centrifugal pump. Journal of Turbomachinery. 1997;**119**(3):506-515

[43] Wernet MP. Development of digital particle imaging velocimetry for use in turbomachinery. Experiments in Fluids. 2000;**28**(2):97-115

[44] Sinha M, Katz J. Quantitative visualization of the flow in a centrifugal pump with diffuser vanes—I: On flow structures and turbulence. Journal of Fluids Engineering. 2000;**122**(1):97-107

[45] Uzol O, Camci C. Aerodynamic loss characteristics of a turbine blade with trailing edge coolant ejection: Part 2—External aerodynamics, total pressure losses, and predictions. Journal of Turbomachinery. 2001;**123**(2):249-257

[46] Wernet MP, Bright MM, Skoch GJ. An investigation of surge in a high-speed centrifugal compressor using digital PIV. Journal of Turbomachinery. 2001;**123**(2):418-428

[47] Uzol O, Chow YC, Katz J, Meneveau C. Experimental investigation of unsteady flow field within a two-stage axial turbomachine using particle image velocimetry. Journal of Turbomachinery. 2002;**124**(4):542-552

[48] Sanders AJ, Papalia J, Fleeter S. Multi-blade row interactions in a transonic axial compressor: Part I—Stator particle image velocimetry (PIV) investigation. Journal of Turbomachinery. 2002;**124**(1):10-18

[49] Estevadeordal J, Gogineni S, Goss L, Copenhaver W, Gorrell S. Study of wake-blade interactions in a transonic compressor using flow visualization and DPIV. Journal of Fluids Engineering. 2002;**124**(1):166-175

[50] Woisetschläger J, Mayrhofer N, Hampel B, Lang H, Sanz W. Laser-optical investigation of turbine wake flow. Experiments in Fluids. 2003;**34**(3):371-378

[51] Uzol O, Chow YC, Katz J, Meneveau C. Average passage flow field and deterministic stresses in the tip and hub regions of a multistage turbomachine. Journal of Turbomachinery. 2003;**125**(4):714-725

[52] Lee SJ, Paik BG, Yoon JH, Lee CM. Three-component velocity field measurements of propeller wake using a stereoscopic PIV technique. Experiments in Fluids. 2004;**36**(4):575-585

[53] Wernet MP, Van Zante D, Strazisar TJ, John WT, Prahst PS. Characterization of the tip clearance flow in an axial compressor using 3-D digital PIV. Experiments in Fluids. 2005;**39**(4):743-753

[54] Yu XJ, Liu BJ. Stereoscopic PIV measurement of unsteady flows in an axial compressor stage. Experimental Thermal and Fluid Science. 2007;**31**(8):1049-1060

[55] Ibaraki S, Matsuo T, Yokoyama T. Investigation of unsteady flow field in a vaned diffuser of a transonic centrifugal compressor. Journal of Turbomachinery. 2007;**129**(4):686-693

[56] Estevadeordal J, Gorrell SE, Copenhaver WW. PIV study of wake-rotor interactions in a transonic compressor at various operating conditions. Journal of Propulsion and Power. 2007;**23**(1):235-242

[57] Voges M, Beversdorff M, Willert C, Krain H. Application of particle image velocimetry to a transonic centrifugal compressor. Experiments in Fluids. 2007;**43**(2-3):371-384

[58] Voges M, Willert CE, Mönig R, Müller MW, Schiffer HP. The challenge of stereo PIV measurements in the tip gap of a transonic compressor rotor with casing treatment. Experiments in Fluids. 2012;**52**(3):581-590

[59] Guillou E, Gancedo M, Gutmark E, Mohamed A. PIV investigation of the flow induced by a passive surge control method in a radial compressor. Experiments in Fluids. 2012;**53**(3):619-635

[60] Gancedo M, Gutmark E, Guillou E. Piv measurements of the flow at the inlet of a turbocharger centrifugal compressor with recirculation casing treatment near the inducer. Experiments in Fluids. 2016;**57**(2):16

[61] Bhattacharya S, Berdanier RA, Vlachos PP, Key NL. A new particle image velocimetry technique for turbomachinery applications. Journal of Turbomachinery. 2016;**138**(12):124501

[62] Ardasheva MM, Nevskii LB, Pervushin GE. Measurement of pressure distribution by means of indicator coatings. Journal of Applied Mechanics and Technical Physics. 1985;**26**(4):469-474

[63] Sabroske K, Rabe D, Williams C. Pressure-sensitive paint investigation

for turbomachinery application. In: ASME 1995 International Gas Turbine and Aeroengine Congress and Exposition; 5 June 1995; American Society of Mechanical Engineers. 1995. p. V001T01A021

[64] Liu T, Torgerson S, Sullivan J, Johnston R, Fleeter S, Liu T, et al. Rotor blade pressure measurement in a high speed axial compressor using pressure and temperature sensitive paints. In: 35th Aerospace Sciences Meeting and Exhibit. 1997. p. 162

[65] Navarra KR. Development of the pressure-sensitive-paint technique for advanced turbomachinery applications [master thesis]. Virginia Tech;

[66] Bencic T. Rotating pressure and temperature measurements on scale-model fans using luminescent paints. In: 34th AIAA/ASME/SAE/ASEE Joint Propulsion Conference and Exhibit; 1 January 1998. p. 3452

[67] Engler RH, Klein C, Trinks O. Pressure sensitive paint systems for pressure distribution measurements in wind tunnels and turbomachines. Measurement Science and Technology. 2000;**11**(7):1077

[68] Navarra KR, Rabe DC, Fonov SD, Goss LP, Hah C. The application of pressure-and temperature-sensitive paints to an advanced compressor. Journal of Turbomachinery. 2001;**123**(4):823-829

[69] Gregory J, Sakaue H, Sullivan J. Unsteady pressure measurements in turbomachinery using porous pressure sensitive paint. In: 40th AIAA Aerospace Sciences Meeting & Exhibit. 2002. p. 84

[70] Lepicovsky J, Bencic T. Use of pressure-sensitive paint for diagnostics in turbomachinery flows with shocks. Experiments in Fluids. 2002;**33**(4):531-538

[71] Gregory J. Porous pressure-sensitive paint for measurement of unsteady pressures in turbomachinery. In: 42nd AIAA Aerospace Sciences Meeting and Exhibit; 5 January 2004. 2004. p. 294

[72] Suryanarayanan A, Ozturk B, Schobeiri MT, Han JC. Film-cooling effectiveness on a rotating turbine platform using pressure sensitive paint technique. Journal of Turbomachinery. 2010;**132**(4):041001

[73] Narzary DP, Liu KC, Rallabandi AP, Han JC. Influence of coolant density on turbine blade film-cooling using pressure sensitive paint technique. Journal of Turbomachinery. 2012;**134**(3):031006

Chapter 4

Development and Control of Generator-Converter Topology for Direct-Drive Wind Turbines

Akanksha Singh

Abstract

In this chapter, a new topology for Direct-Drive Wind Turbines (DDWTs) with a low-voltage generator design is presented in order to eliminate the required dc-bus capacitors or dc-link inductors. In the presented topology, the grid-side converter is replaced by a boost Current Source Inverter (CSI) therefore removing the need for the dc-bus electrolytic capacitors which results in increasing the system lifetime. In the developed topology, the synchronous inductance of the generator is utilized. This facilitates the elimination of the intrinsically required dc-link inductor in the CSI which further contributes to a reduction in the overall system weight and size. The boost CSI is capable of converting a low dc voltage to a higher line-to-line voltage. This results in the implementation of a low-voltage generator for DDWTs. The feasibility of the presented low-voltage generator is investigated through Finite Element (FE) computations. In this chapter, a modified 1.5 MW low-voltage generator for the proposed topology is compared with an existing 1.5 MW Permanent Magnet (PM) synchronous generator for DDWTs. The feasibility of the presented topology of generator-converter for DDWTs is verified through simulations and laboratory tests. Furthermore, the controls developed for the developed wind turbine topology is also presented in this chapter.

Keywords: boost CSI, direct-drive wind turbine, finite element analysis, permanent magnet synchronous generator, wind turbine topologies

1. Introduction

Wind power is one of the fastest growing energies and the global capacity has increased to 433 GW by the end of 2015 [1, 2]. The two most commonly used topologies for wind turbines are based on the Doubly Fed Induction Generator (DFIG) or Permanent Magnet Synchronous Generator (PMSG) [3]. In the DFIG configuration, the stator of the generator is directly connected to the grid while the rotor does not require to rotate at the fixed synchronous speed. The rotor is connected to the turbine shaft through a gearbox. In the PMSG-based wind turbines, the speed of the rotating magnetic field and the rotor is the same and therefore, it is connected to the grid through power converters. The PMSG-based wind turbines require a power electronics interface to be connected to the grid which provides the flexibility of using these turbines with or without gearbox between the turbine shaft and the generator shaft [3]. It has been shown that the gearboxes cause more downtime than any other component in a wind turbine [3–5]. It is worth noting that the gearboxes are responsible for 10% of the wind turbine failures which result in about 20% of the total wind turbine downtime [4–6]. Recent investigations reveal that gearboxes in wind turbines, which were supposed to last 20 years, might fail in 7–10 years [7, 8]. The Direct-Drive Wind Turbines (DDWTs) do not have a gearbox between the turbine rotor and the generator shaft. There is a definite trend toward DDWTs as is predicted in the research papers and trade articles but there are some major concerns that must be overcome in order to achieve higher market penetration [1, 9]. The power electronics interface in a DDWT is rated for full power transfer and it consists of an active rectifier and a grid-side converter, connected through a capacitor bank forming the dc-bus. The power electronics interface is one of the most vulnerable components in wind turbines [4, 5, 7]. A significant percentage of these failures have been attributed to the dc-bus electrolytic capacitors [10, 11]. Many methods which try to determine the remaining lifespan of electrolytic capacitors exist and are utilized for scheduled maintenance planning [5]. These methods do not serve to extend the life of the capacitors and contribute to an increase in the system downtime [7]. The other major concern which prevents the proliferation of DDWTs is the large size of PMSGs [12, 13]. The large size of the generator is a result of the generator shaft rotating at the same speed as the turbine rotor shaft. The low angular speed in PMSGs increases the required pole surface and the number of poles, which results in a high cost and volume of the generator.

In the most commonly used DDWT configuration the power electronics interface comprises of back-to-back Voltage Source Converters (VSCs) connected through electrolytic capacitors [14]. In this paper, the grid-side VSC is replaced by the boost Current Source Inverter (CSI). The proposed boost CSI topology of the converter results in the elimination of the dc-bus electrolytic capacitors, which are one of the frequently failing components in the existing topologies of direct-drive wind turbines without any supplementary component at the dc-bus [10, 11, 15]. The boost CSI converts the low dc voltage of the generator rectifier to the higher voltage level which facilitates the implementation of low voltage PMSG with lower weight and volume [16]. In addition to the introduction, this chapter has five more sections. In Section 2, the existing wind turbine topologies are reviewed, and the topology of the developed system is introduced. Section 3 presents the feasibility of a 1.5 MW low-voltage PMSG for the presented system. This section also presents a comparison between the generator for the developed system and an existing 1.5 MW PMSG for a DDWT using Finite Element (FE) computations. The controls developed for the new DDWT topology are presented in Section 4. The simulation results demonstrating the feasibility of the developed system are presented in Section 5. A laboratory scale setup is used to experimentally confirm the feasibility of the proposed topology in Section 6. Section 7 presents the conclusions of the chapter.

2. Wind turbine topologies

In this section, several wind turbine topologies are reviewed prior to introducing the inductorless current source generator-converter topology.

2.1 Indirect drive wind turbine topologies

In the majority of wind turbine topologies, the windmill rotor shaft is connected to the generator through a gearbox [14, 17]. The indirect drive wind turbines can be assembled using various types of generators and converters. Most commonly used indirect wind turbine utilizes the DFIG configuration of the drivetrain. In this configuration, the rotor is connected to the grid through a back-to-back converter while the stator windings are directly connected to the grid and. This back-to-back converter includes of two bidirectional converters connected at the dc-bus, which formed by a capacitor bank. In this configuration, the major part of the power injected into the grid is through the stator of the generator and the back-to-back converters transfer only a fraction of the power produced by the wind turbine into the grid. The PMSG with the same back-to-back converter topology is another common indirect drive wind turbine topology. Unlike in the DFIG configuration, the converter is rated for the full power.

2.2 Direct drive wind turbine (DDWT) topologies

The DDWT is a relatively new topology and has the turbine rotor shaft directly connected to the generator shaft without any gearbox. **Figure 1** shows commonly used topologies for DDWTs using the PMSG. **Figure 1(a)** displays the power electronics interface which consists of a diode bridge rectifier, dc-dc boost converter, and grid-tied VSC [17, 18]. **Figure 1(b)** shows another DDWT topology with power electronics interface consisting of grid-tied back-to-back Current Source Converters (CSCs) connected through a dc-link inductor [19, 20]. **Figure 1(c)** demonstrates the most frequently used configuration of DDWTs, comprising of a PMSG connected

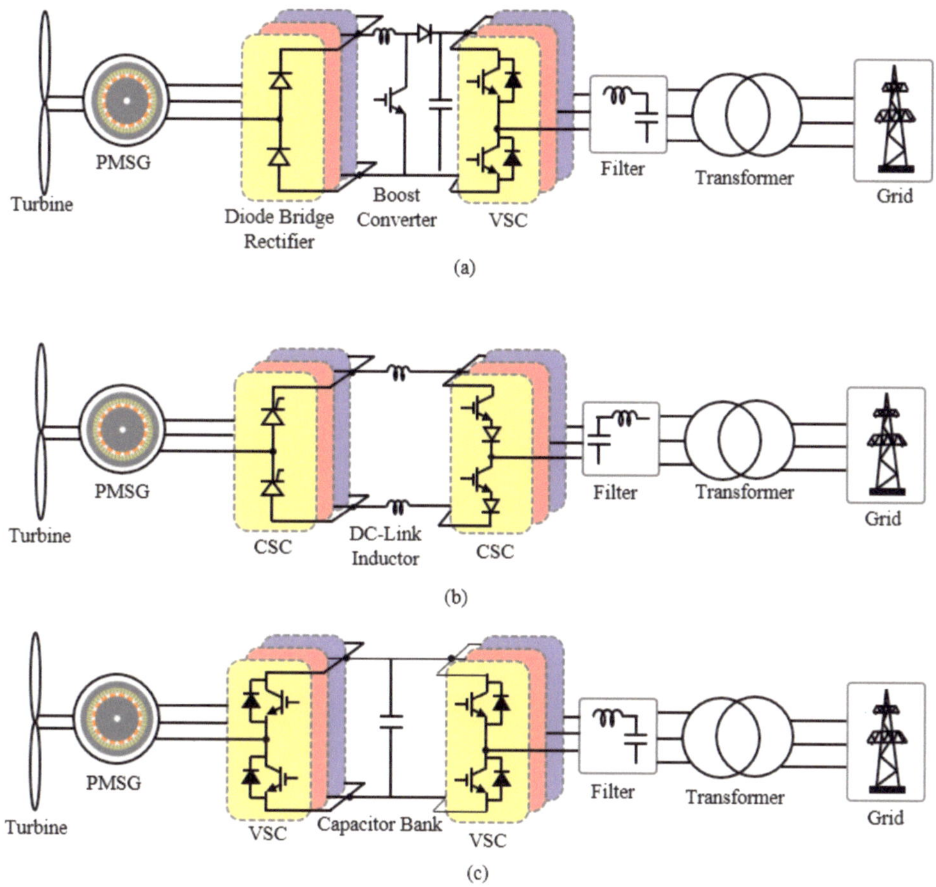

Figure 1.
Common DDWT topologies with (a) PMSG connected to diode-bridge, boost converter and VSI, (b) PMSG connected to back-to-back CSCs, and (c) PMSG connected to back-to-back VSCs.

to the grid through two VSCs [17, 21, 22]. The dc-bus between the VSCs is formed using electrolytic capacitors in order to regulate and stabilize the dc-bus voltage. In some cases, multilevel converters are used to form the power electronics interface in the drivetrain. The topology of the multilevel converters can be an H-bridge back-to-back converter or a neutral-point-clamped back-to-back converter [17, 22]. Most wind turbine topologies either have a dc-bus formed by electrolytic capacitors or a dc-link formed by inductors. The electrolytic capacitors are one of the most failure-prone components that adversely impact the system reliability [5]. The failure of these capacitors has a significant impact on the maintenance cost especially in the case of offshore wind turbines [13, 23]. The dc-link inductor is bulky and adds to the system loss reducing the system's efficiency [19]. In some cases, the grid side converter is located away from the wind turbine and the transmission cable is used to realize the dc-link inductor [20]. Additionally, the lack of gearbox in the DDWT increases the size of the PMSG and the capital cost of the overall system [13, 23–25]. In the following, the proposed DDWT topology is described.

2.3 Proposed DDWT topology

The proposed topology for the DDWT is presented in **Figure 2**. In this subsection, different parts of the proposed topology are explained. First, the power electronics interface topology is presented, and then, the PMSG design flexibility provided by implementing the boost CSI is described.

2.3.1 The power electronics interface

The power electronics interface in the proposed topology is a back-to-back converter made up of a three-phase VSC and a boost CSI, as shown in **Figure 2**. The boost CSI is equipped by Reverse Blocking IGBTs (RB-IGBT), and no dc-bus capacitor or dc-link inductor exists in the proposed DDWT. As a result of avoiding the dc-bus capacitor, the system mean-time-between-failure can significantly be improved in comparison with the conventional VSC-based systems [26]. While any CSI requires an inductor at the dc-link, the proposed DDWT eliminates the dc-link inductor by utilizing the generator synchronous inductance, L_s [15]. In order to achieve a Total Harmonic Distortion (THD) of the current waveforms at the inverter output acceptable under the IEEE 1547–2018 interconnection standards, the operation of the boost CSI must always be in Continuous Conduction Mode (CCM). The minimum dc-link inductance, L_{dc} required to keep the inverter in CCM has been derived in [27]. However, the synchronous inductance, L_s, will be less than the L_{dc}, i.e. $L_s = \sqrt{3/2}\, L_{dc}$. This is because Thevenin's equivalent inductance of the generator-converter from the dc-bus is almost $(3/2)\, L_s$ and $I_{rms} \approx \sqrt{2/3}\, I_{dc}$. Also, a traditional CSI operated using space vector PWM switching provides a maximum

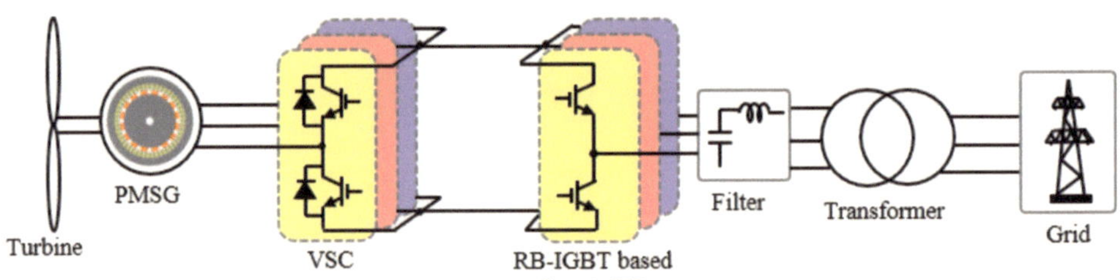

Figure 2.
The proposed wind turbine system topology with PMSG connected to a VSC and the boost inverter equipped by RB-IGBTs.

boost ratio, V_{LLrms}/V_{dc}, of about 1.2 [28]. However, the boost CSI is modulated using the Phasor Pulse Width Modulation (PPWM) switching technique, providing a boost ratio more than three [29–31]. As a result, a low voltage PMSG design will be possible for the proposed DDWT. The advantages gained by using a low voltage generator are discussed in the next subsection.

2.3.2 The permanent magnet generator

The low dc-side voltage required by the boost CSI enables a low output ac-voltage from the PMSG. In a traditional DDWT topology, the voltage at the generator and the grid should have almost the same rms value [32, 33]. However, if the developed topology is utilized, then, for example, 120 V_{LLrms} from the PM generator is sufficient to generate an rms line-to-line voltage of 480 V at the converter output. As described in [15], consider the emf equation $E \propto N_{ph} p\, \omega_m \phi_p$, where, E is the peak induced voltage per phase, N_{ph} is the number of winding turns per phase, ω_m is the mechanical angular velocity, p is the total number of poles, and ϕ_p is the maximum magnetic flux per pole, [33]. Here ω_m is dependent on the wind speed, and the fundamental magnetic flux per pole is obtained from $\phi_p \cong (4/\pi) B_m (2\pi/p)(D/2)l = 4Dl B_m/p$, where D is the mid-airgap diameter, B_m is the flux density, and l is the stack length. Accordingly, for a given angular velocity, permanent magnet material, and air-gap height, l_g one can say $E \propto N_{ph} Dl$, which means that a generator with a lower output voltage requires a lower value of $N_{ph} Dl$. On the other hand, the minimum synchronous inductance, required for the boost CSI to operate in CCM, restricts the maximum number of poles, since $L_s \propto (Dl/l_g)\,(N_{ph}/p)^2$ [33]. Hence, the desired values of the generator output voltage, $(E^2 - (\omega_s L_s I)^2)^{1/2} \cong E$ and synchronous inductance, L_s, are herein used as the PMSG design inputs for a given rated power.

3. Generator design and comparison

In this section, a 1.5 MW PMSG is designed for the proposed DDWT topology and then the design is compared with an existing 1.5 MW PMSG [34].

3.1 New generator design

The generator is designed using GenAC toolbox of MagneForce Finite Element (FE) software. The designed machine is a low-speed, concentrated overlapping tooth coil double layered winding, three-phase PMSG rated at 1.5 MW. The generator is designed to be radial flux, cylindrical rotor, rotor surface mounted pole machine so that it can be compared with the existing 1.5 MW PMSG [34]. For the sake of comparison, both the existing and new 1.5 MW systems have the rated speed of 19.65 rpm, and the same generator airgap of 0.6 mm. Also, the line-to-line voltage of the new generator is 450 V, while the existing generator line-to-line voltage is 770 V, see **Table 1**.

Figure 3 shows the cross-sectional view of the new generator, including the stator slot and tooth dimensions in details. Moreover, the insulation thickness in each slot is 3.5 mm. The overall slot-fill is about 62%. Also, the rotor shaft bore diameter is 700 mm. The stator bore diameter is 3500 mm, the stator length is 600 mm, and the airgap length, l_g, is 0.6 mm. The PM material is sintered Neodymium Iron Boron (NdFeB) with the maximum residual flux density of B_r = 1.2 T, and maximum coercive force of 24 k Oersted (i.e. H_c =1910 kA/m). The permanent magnet pole dimension is as 260 × 100 mm^2 with a pole arc to pole pitch ratio of 0.8. **Figure 4** demonstrates the electromagnetic flux and the flux density distribution over the

Parameter	Existing generator	Designed (new) generator
Power rating	1.5 MW	1.5 MW
Speed	19.65 rpm	19.65 rpm
Air gap length	0.6 mm	0.6 mm
Line-to-line voltage	770 V	450 V

Table 1.
The existing and new generator input parameters.

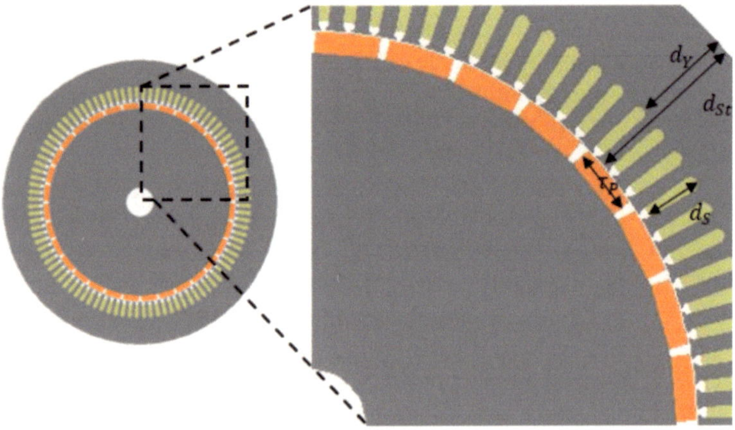

$d_Y = 350$mm, $d_{St} = 420$mm, $\tau_P = 260$mm, $d_S = 50$mm, air-gap length $l_g = 0.6$mm, slot opening width = 10mm, slot width = 65 mm

Figure 3.
Cross sectional view of the new design 1.5 MW PMSG.

PMSG at 1 *p.u.* load. As can be seen in **Figure 4**, the generator is not saturated at rated load. The maximum flux density is observed at the stator tooth, and the maximum airgap flux density measured using FE computations is about 1.15 T. Using circuit simulation for the new design, ~20 mH is needed to keep the dc-link current in CCM for different load levels.

3.2 Generator parameter comparison

In this subsection, the designed PMSG is compared with an existing 1.5 MW PM generator for DDWT [34]. **Table 2** shows the differences between the new and existing design parameters. In order to compare the two generators, the existing generator design is duplicated using the FE software. The design parameters for the existing generator are obtained from [34]. The comparison of the generator no-load line-to-line voltage, and the flux linkage with respect to rotor electrical position is presented in **Figure 5**. The no-load line-to-line voltage waveforms are shown in **Figure 5(a)**. The flux linkage of phase-A of the generator versus the rotor electrical position is presented in **Figure 5(b)**. As can be seen in **Figure 5(b)** that there is minimal distortion in the flux linkage, and it is sinusoidal for the generator designed. Furthermore, **Figure 5(b)** shows that the flux linkage is higher for the new design than the existing generator, as the new design has higher per phase inductance than that of the existing generator. It should be noted that even though the flux is higher in the new generator design, the maximum flux density is less than that permissible for the core, i.e. 1.2 T. As shown in **Table 2**, the number of

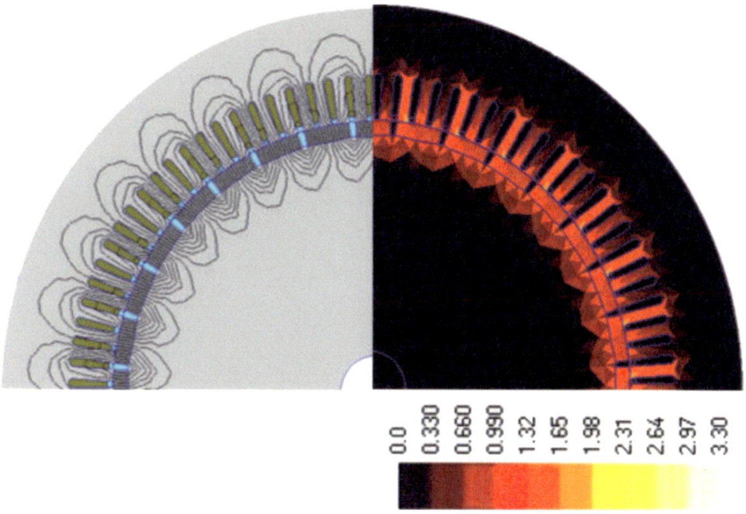

Figure 4.
Flux and flux density distribution (T) over the cross-section of the designed PMSG at full load.

Parameter	Existing generator	Designed (new) generator
No. of poles, p	78	32
Airgap flux density, B_{max}	1.01 T	1.15 T
Winding current density, J	4 A/mm^2	6 A/mm^2
Linear current density	31.08 kA/m	36.60 kA/m
Stator winding factor, k_w	0.942	0.951
Stator q-axis inductance, L_q	12.20 mH	20.16 mH
Stator d-axis inductance, L_d	12.19 mH	20.11 mH
Stator inner diameter	4462 mm	3500 mm
Stack length	500 mm	600 mm
Number of slots	234	96
Winding turns per phase	56	30
Copper weight	384.4 kg	401.3 kg
Core weight (stator)	21,552 kg	19,429 kg
Rotor weight	40,990 kg	35,870 kg

Table 2.
Comparison of the designed vs. existing generator.

poles in the existing generator is 78 which has been reduced to less than half in the new design to 32 poles. Also, each pole volume is 600 × 260 × 100 mm^3 in the new design, whereas, in the existing PMSG, each pole volume is 500 × 142 × 100 mm^3. Thus, the total permanent magnet material is reduced by 9.8% in the new design. The reduction in the permanent magnet material reduces the dependency on the imported and highly unstable market of NdFeB materials.

The FE measured linear current density for the new design is 36.6 kA/m at the pole surface, as shown in **Table 2**, while the NdFeB coercive force is 1910 kA/m. In **Table 3**, the demagnetization of PM poles is computed for different output power factors and the results are presented. The computation of demagnetization is done using the MagneForce FE software. The software calculates the total flux with

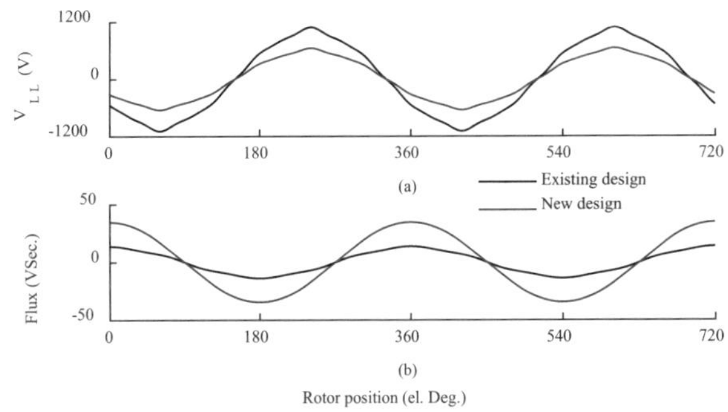

Figure 5.
Generator (a) line-to-line voltage, and (b) phase flux linkage vs. rotor electrical position obtained through FE computations for the existing and new design of the generators.

	pf = 0.95	0.85	0.75
		Demagnetization (%)	
Existing design	0.080	0.075	0.078
New design	0.084	0.080	0.081

Table 3.
Percent demagnetization of pm poles for varying output power factor and s = 1 p.u. using FE computations.

virgin (non-demagnetized) magnets. It then runs the simulation and keeps track of the worst demagnetization that each element within the rotor pole experiences throughout a complete ac cycle. The FE software uses the demagnetized elements to re-construct the magnet, and then, calculates the total flux from this demagnetized magnet. Finally, the software calculates the percentage drop in total flux from the virgin magnets to the demagnetized magnets and reports the difference as the amount of demagnetization. It should be noted that the demagnetization analysis is based on loading of the generators and total flux drop, and it does not take into account the thermal models for the generators.

A normalized comparison of the copper and core losses, rotor and stator weights, and the volume of the two generators is presented in **Figure 6**. In **Table 4**, the efficiency, as well as core and copper losses, are provided for different power factors and a constant apparent power, i.e. S =1 p.u. The base values for the per-unit (p.u.) calculations are S_{base} =1.5 MVA, and V_{base} =690 V. As can be observed, the copper loss in the new generator increases. This is due to the higher current level in the low-voltage generator. Although the new design has lower number of turns in the generator winding – lower overall stator resistance – the line current is higher, which results in higher copper loss in the new low voltage generator. Nevertheless, the core loss for the designed low voltage generator decreases by about 9.5%. This reduction in the core loss in the new designed generator is due to the reduction in the number of poles in the new designed generator, which results in a substantial decrease in the generator output voltage and frequency, see **Table 2**. It can be observed that the generator efficiency is not compromised with full load efficiency of the new and existing generator being 94.4 and 93.8%, respectively, while the stator and rotor dimensions are significantly decreased. The no-load core-losses for the existing and new generator designs are computed to be 0.040 p.u. and 0.0351 p.u., respectively. The change in core losses from no-load to full load can be attributed to the armature reaction, see **Table 4**. It should be noted that the losses and efficiency data presented in in **Table 4**, and **Figure 6** are

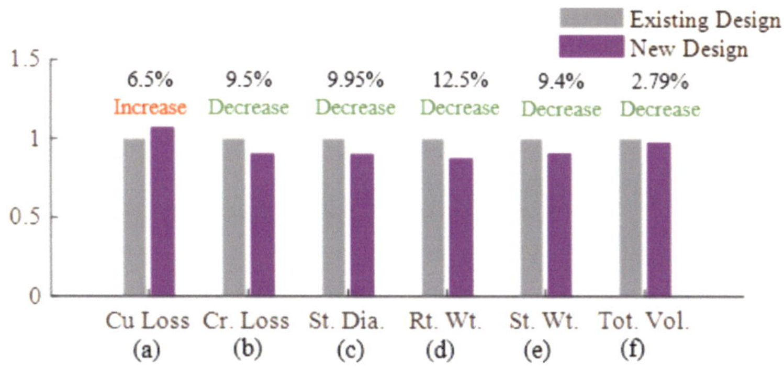

Figure 6.
Comparison of normalized (a) copper loss, (b) core loss, (c) stator diameter, (d) rotor weight, (e) stator weight, and (f) volume of the existing and designed PM generator for DDWT using FE computation.

	pf =	**0.95**	**0.85**	**0.75**
Copper Loss (*pu.*)	Existing	0.0245	0.0245	0.0250
	New	0.0268	0.0269	0.0278
Core Loss (*pu.*)	Existing	0.0405	0.0405	0.0405
	New	0.0353	0.0354	0.0354
Efficiency (%)	Existing	93.50	93.49	93.45
	New	93.79	93.77	93.68

Table 4.
Core loss, copper loss, and efficiency for varying output power factor and S = 1 p.u. using FE computations.

obtained using the FE models of the PMSGs. A comparison of the rotor and stator weights, respectively, for the two generators is presented in **Figure 6**. The rotor weight is the cumulative weight of the NdFeB PM poles and the rotor core. Similarly, the stator weight presented in the combined weight of the armature windings and the stator core. The stator and rotor weight reductions for the new generator are ~9.4 and 12.5%, respectively. Additionally, the overall volume of the new generator is reduced by ~ 2.79% than the existing design. It can be observed that the decrease in the stator diameter (which is about 10%) does not render to a comparable reduction in the overall volume of the new generator. This is due to the higher stack length of the new generator. The designed generator has a stack length to diameter (l/D) ratio of 0.13 as compared to the existing generator design with l/D ratio of 0.11. The decrease in the generator volume and weight along with the elimination the failure prone dc-bus capacitor and circumventing the dc-link inductor needed in CSIs will further render into decreased capital cost of the overall system. The simulation and experimental results evaluating and validating the performance of the proposed topology of the DDWT are presented in the following sections.

4. System controls

In this section, first, the control technique implemented for the developed topology is briefly explained. The developed system is controlled with generator side VSC being modulated for power transfer to the grid and the boost inverter modulated to stabilize the input dc-voltage.

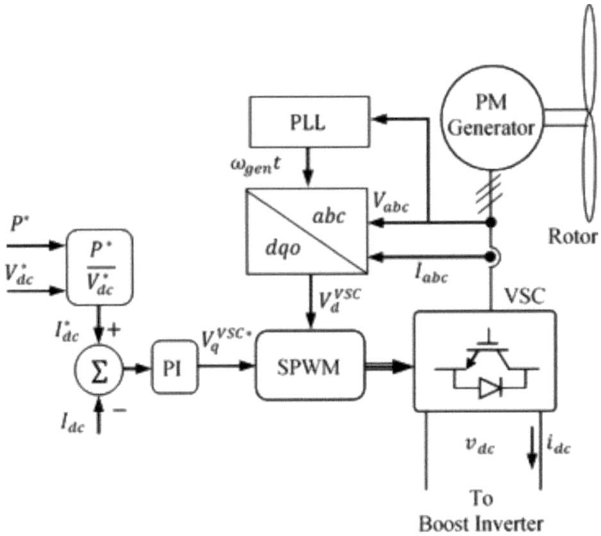

Figure 7.
Block diagram of controller for the VSC [26].

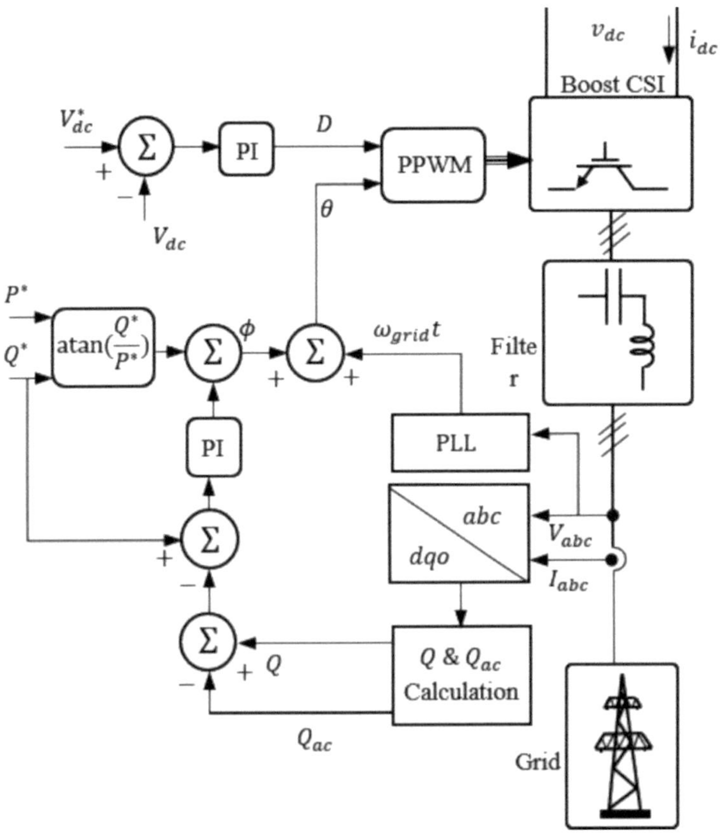

Figure 8.
Block diagram of controller for the single-stage boost inverter [26].

The power electronics interface proposed for the DDWT drivetrain is controlled to regulate the power injected to the grid and to maintain the average dc-side voltage. The generator side VSC is modulated to inject desired power to the dc-side.

Figure 7 shows the block diagram of the controller for VSC. The generator line-to-line voltages and dc-side voltage and current are the measured feedback received by this controller. The ac side signals are then converted to the *dqo* reference frame. The desired active power on the grid side, P^* is compared to the measured power on the dc side. The error is then mitigated by using a Proportional-Integral (PI) controller which then generates the desired q-axis converter voltage, V_q^{VSC*}. The generated q- axis voltage and the measured d-axis voltages serve as base to generate the switching signal for the VSC. The controller is equipped with one of the existing MPPT techniques, e.g. $P^* = K_{opt}\omega_R^3$, [36, 37].

Figure 8 shows the block diagram of the controller for the boost CSI. The controls for this converter help regulate the dc-side voltage and to modulate the reactive power injected into the grid. Thus, the controller for the VSC helps modulate the active power injected into the grid and the controller for the boost CSI modulates the reactive power into the grid. The inputs to the boost CSI controller are the measured grid-side line-to-line voltages, line currents and dc-side. The three-phase signals are converted to the *dqo* reference frame. The *dqo* quantities are used to compute the reactive power injected to the grid, which is compared to the reference reactive power. The error is then mitigated using a PI controller which generates the reference angle, θ for the boost inverter. Similarly, the dc-side voltage, V_{dc} (where, $V_{dc} = \langle v_{dc} \rangle$) is compared to the desired dc-side voltage, V_{dc}^* and a PI controller is used to generate the modulation index, D for the implementing the PPWM switching technique [26]. The details on the PPWM technique has been provided in [29–31].

5. Simulation results

This section presents the simulation results for the evaluation of the feasibility and performance of the developed topology of the DDWT. The software platform used for these evaluations is MATLAB/Simulink environment using SimPowerSystems toolbox. The simulated system uses the FE design of the generator as presented in Section 3. The switching frequency of 6 kHz is used for both the VSC and the boost CSI. The switching methodology for the VSC is the Sine PWM (SPWM) technique while the boost CSI employs the PPWM switching technique [29, 30]. As described in the previous section, the VSC controller modulates the active power transfer injected into the grid and the boost CSI controller regulates the dc-side voltage and controls the reactive power injected into the grid. A detailed discussion on these controllers and their performance is presented in [26]. The output filter parameters for the boost CSI are C_f = 20 μF and L_f = 5 mH.

The quality of the output waveforms of the proposed system is demonstrated in **Figures 9** and **10**. **Figure 10** presents the steady-state waveforms of generator line-to-line voltage and line current when the generator speed is 18.75 rpm. The power converters were modulated to inject 1 MW (0.67 p.u.) P into the grid at unity power factor (UPF). The boost CSI controller was regulating the dc-side voltage at 500 V. **Figure 10(a)** and **(b)** show that the THD of the generator output line current is ~1.2%, which further ensures less harmonics transferred to the generator side and eliminates the low-frequency torque ripple on the generator shaft. **Figure 9** presents the waveforms of the boost CSI output line-to-line voltage at the point of common coupling, and line current injected into the grid. The grid line current THD is computed to be ~3.3%, which is complaint of the IEEE 1547–2018 interconnection standards [38].

Figure 11 presents the dynamic behavior of the proposed DDWT system. The generator speed, three-phase line-to-line voltage waveforms at the point of common coupling (PCC), three-phase generator line current waveforms, and the dc-side current are shown in **Figure 11**. The first event occurs at t = 1*s*, when the generator speed

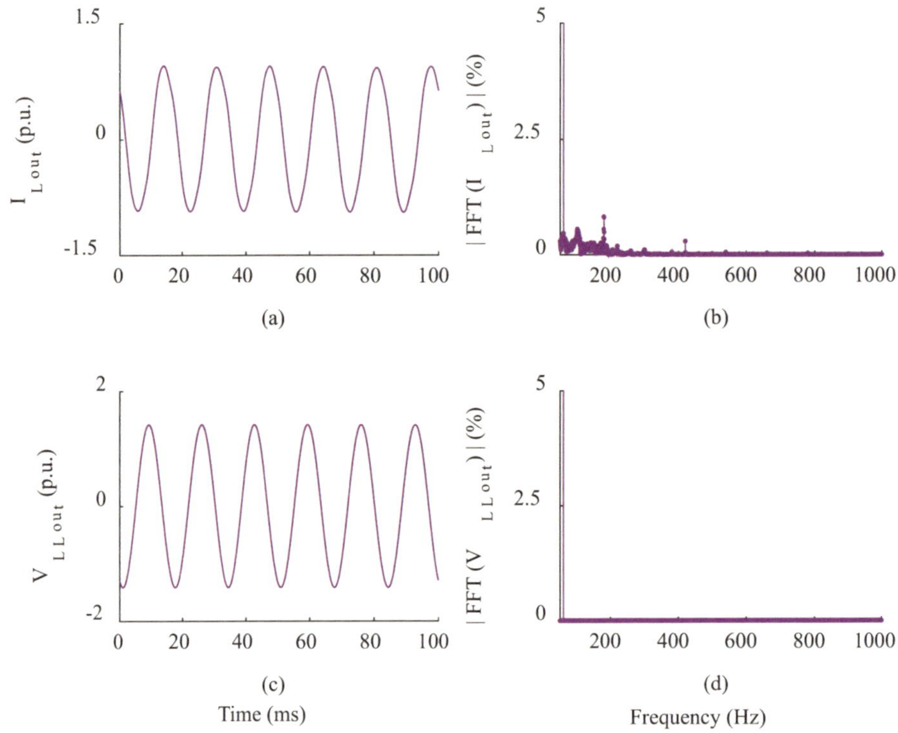

Figure 9.
Simulated results of inverter output (a) line current waveform, (b) line current FFT analysis, (c) line-to-line voltage waveform, and (d) line-to-line voltage FFT analysis for grid-tied system the generator speed is 18.75 rpm and the system is injecting 0.67p.u. Active power into the grid [35].

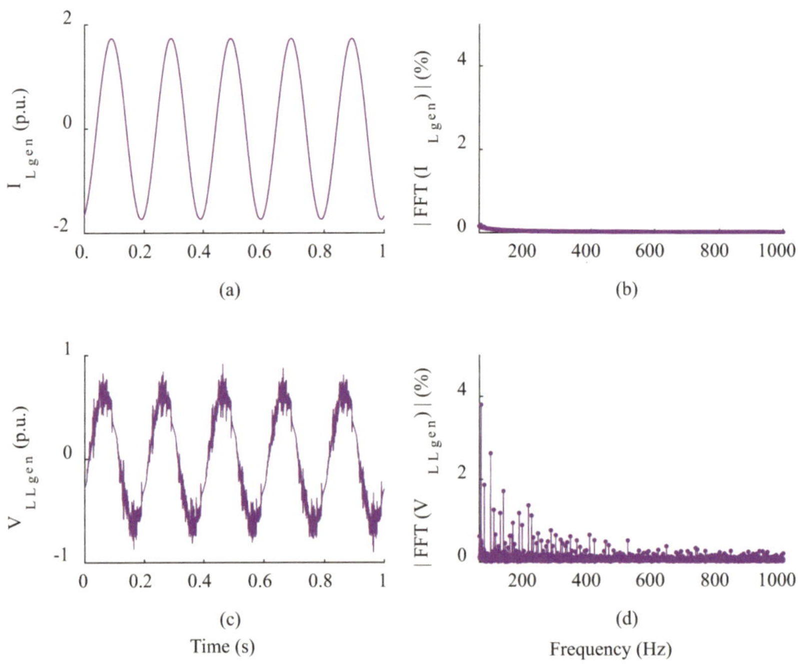

Figure 10.
Simulated results of generator output (a) line current waveform, (b) line current FFT analysis, (c) line-to-line voltage waveform, and (c) line-to-line voltage FFT analysis, for grid-tied system when the generator speed is 18.75 rpm and it is supplying [35].

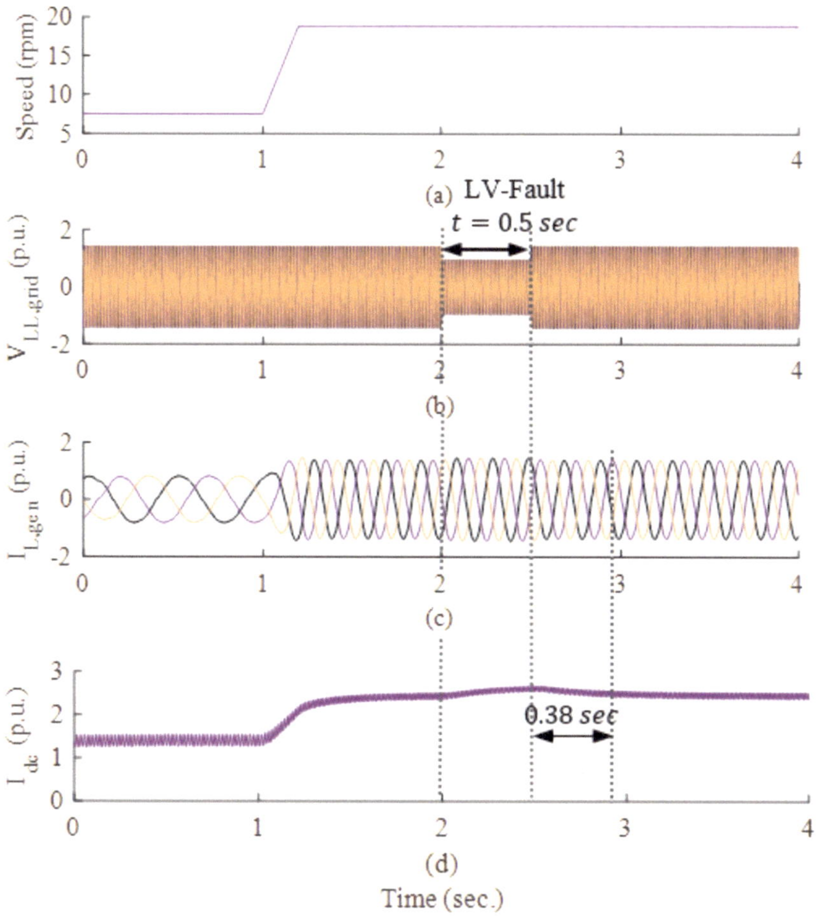

Figure 11.
Plot of (a)PMSG speed, (b) three-phase line-to-line voltage at the PCC, (c) three-phase generator line current, and (d) dc-side current showing system behavior for varying wind speed and three-phase low-voltage fault at the grid [35].

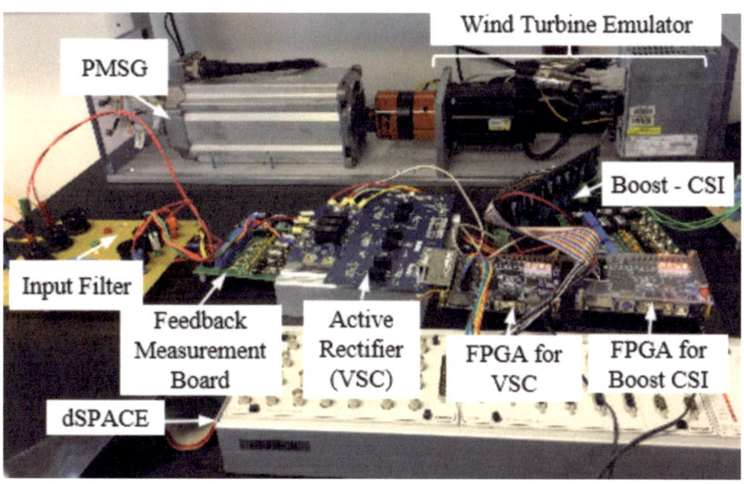

Figure 12.
Laboratory scale 1.5 kW, setup for the proposed DDWT topology [35].

changes gradually from 7.5 rpm to 19.65 rpm. The sudden speed changes shown in **Figure 11** are not realistic and is just used as a case to evaluate the robustness of the system controls. **Figure 11(c)** and **(d)** show that the system easily tracks the maximum available wind power. The fault ride-through capability of the system is also evaluated. This is done by simulating a three-phase low-voltage fault on the inverter terminals at t = 2 s, where the grid voltage decreases by 35% and the fault is cleared after 0.5s [39]. The developed system rides-through this low-voltage seamlessly and starts normal operation again in ~0.38s after the clearing of the fault.

6. Experimental evaluation results

The feasibility of the developed DDWT topology is evaluated using experimental results from 1.5 kW, 240V_{LL} laboratory scale prototype, in this section (see **Figure 12**). In this setup, a motor-drive system is used as the wind turbine emulator that runs a commercially available PMSG, which feeds a three-phase 50Ω resistive load through the proposed back-to-back converter and a CL filter. The filter values are C_f = 20 μF and L_f = 5 mH, when f_s = 6 kHz for the boost CSI. For the first test case to evaluate if the generator synchronous inductance can be utilized in place of the dc-link inductor, a diode bridge rectifier is used as the generator side. The waveforms of the generator line current, generator line-to-line voltage, load line current, and load line-to-line voltage are shown in **Figure 13**. The PMSG produces rms line-to-line voltage of 45 V at 30 Hz. This is boosted to an output load line-to-line voltage of 208 V at 60 Hz by the boost CSI. The generator current and load current

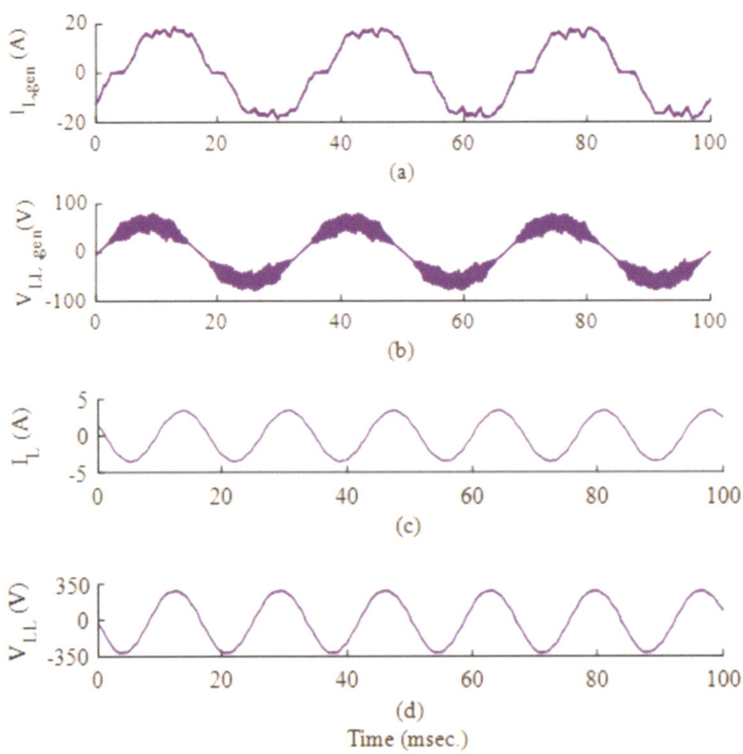

Figure 13.
Experimentally obtained waveforms of the (a) generator line current, (b) generator line-to-line voltage, (c) load line current, and (d) load line-to-line voltage when the generator is rotating at 450 rpm (30 Hz), the output is regulated at 208 V, and the generator side converter is a diode bridge rectifier [35].

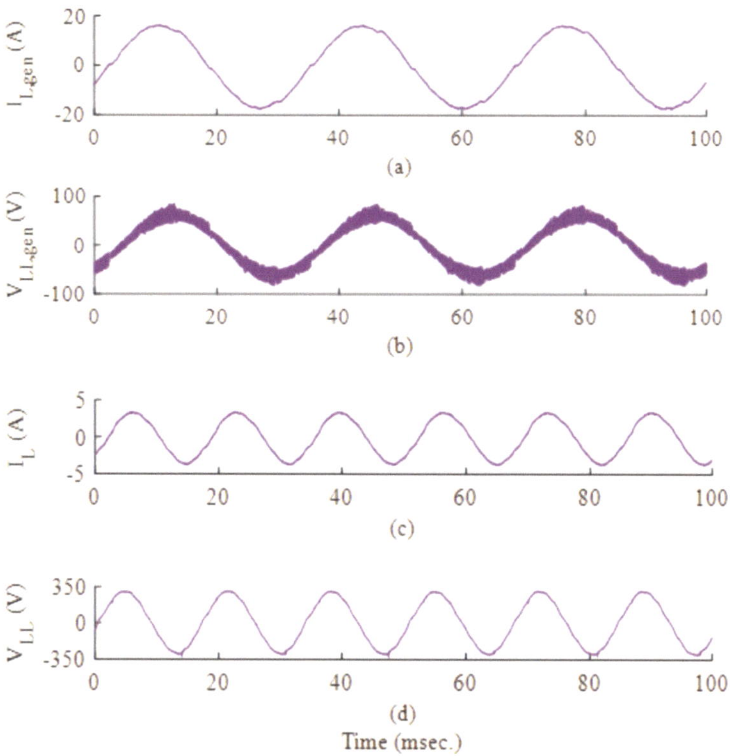

Figure 14.
Experimentally obtained waveforms of the (a) generator line current, (b) generator line-to-line voltage, (c) load line current, and (d) load line-to-line voltage when the generator is rotating at 450 rpm (30 Hz) and the output is regulated at 208 V with the generator side converter as VSC [35].

THDs are measured to be 33.1 and 3.0%, respectively. For the next test case, the generator side converter employed is a VSC. **Figure 14** shows the waveforms of the same quantities for this test case. The switching frequency for the VSC is $f_s = 25$ kHz. The average dc side voltage was maintained at 70 V. The generator current and load current THDs are measured to be 2.9 and 2.8%, respectively The evaluation results presented in **Figures 13** and **14** demonstrate the capability of the developed systems to utilize the generator synchronous inductance, while still maintain the quality of the system output.

7. Conclusion

In this chapter, an inductorless generator-converter topology has been presented for DDWTs. This paper also presents a new topology of power electronics interface for the DDWTs. The grid side VSC used in a conventional DDWT has been replaced by a boost CSI. The dc-link inductor inherently required in a CSI topology has been eliminated by using the synchronous inductance of the PMSG. This topology has further facilitated the elimination of vulnerable dc-bus electrolytic capacitors, thereby increasing the overall system reliability. Furthermore, less frequent failures of the capacitors will lead to increase in system availability, resulting in substantial decrease in the maintenance costs. In this chapter, FE analysis is used to design and evaluated a 1.5 MW generator for the proposed system. The FE analysis is also used to compare the designed and an actual generator. The advantages of using the proposed topology have been demonstrated through reduction in generator weight, volume, and amount of PM material. The controls for the proposed system have also been discussed in this chapter. The feasibility and performance of the proposed DDWT system has been validated through simulation results and experimental results from a laboratory scale prototype, in this chapter.

Author details

Akanksha Singh
National Renewable Energy Laboratory, Golden, USA

*Address all correspondence to: akanksha.singh@nrel.gov

IntechOpen

© 2019 The Author(s). Licensee IntechOpen. This chapter is distributed under the terms of the Creative Commons Attribution License (http://creativecommons.org/licenses/by/3.0), which permits unrestricted use, distribution, and reproduction in any medium, provided the original work is properly cited. CC BY

References

[1] Global Wind Energy Council Report. [Online] Available from: http://www.gwec.net/global-figures/wind-in-numbers/

[2] AWEA U.S. Wind Industry Market Reports for 2015 [Online]. Available from: http://www.awea.org

[3] Manwell J, Mcgowan J, Rogers A. Wind Energy Explained: Theory, Design, and Applications. Second ed. United Kingdom: Wiley; 2011

[4] Ribrant J, Bertling L. Survey of failures in wind power systems with focus on swedish wind power plants during 1997-2005. IEEE Transactions on Energy Conversion. 2007;**22**(1):167-173

[5] Daneshi-Far Z, Capolino GA, Henao H. Review of failures and condition monitoring in wind turbine generators. In: International Conference on Electrical Machines. 2010

[6] Tavner P, Bussel G, Spinato F. Machine and converter reliabilities in wind turbines. In: IEEE International Conference on Power Electronics, Machines and Drives. Dublin, Ireland; 2006

[7] Polinder H, Van Der Pijl FFA, de Vilder GJ, Tavner P. Comparison of direct-drive and geared generator concepts for wind turbines. IEEE Transactions on Energy Conversion. 2006;**21**(3):725-733

[8] Spinato F, Tavner PJ, van Bussel GJW, Koutoulakos E. Reliability of wind turbine subassemblies. IET Renewable Power Generation. 2009;**3**(4):387-401

[9] Liserre M, Cardenas R, Molinas M, Rodriguez J. Overview of multi-MW wind turbines and wind parks. IEEE Transactions on Industrial Electronics. 2011;**58**(4):1081-1095

[10] Wang H, Liserre M, Blaabjerg F. Toward reliable power electronics: Challenges, design tools, and opportunities. IEEE Industrial Electronics Magazine. 2013;**7**(2):17-26

[11] Chen J, Chen W, Li J, Zhang X, Sun P. Lifetime assessment of DC link electrolytic capacitor of wind power converter based on operational condition. In: 2016 IEEE International Conference on High Voltage Engineering and Application (ICHVE), Chengdu. 2016. pp. 1-4

[12] Leban K, Ritchie E, Argeseanu A. Design preliminaries for direct drive under water wind turbine generator. In: IEEE International Conference on Electrical Machines. Marseille, France; 2012

[13] Mcmillan D, Ault GW. Techno-economic comparison of operational aspects for direct drive and gearbox-driven wind turbines. IEEE Transactions on Energy Conversion. 2010;**25**(1):191-198

[14] Chen Z, Guerrero JM, Blaabjerg F. A review of the state of the art of power electronics for wind turbines. IEEE Transactions on Power Electronics. 2009;**24**(8):1859-1875

[15] Singh A, Mirafzal B. Indirect boost matrix converter and low-voltage generator for direct drive wind turbines. The Journal of Engineering. 2018;**2018**(1, 1):10-16

[16] Singh A, Mirafzal B. A generator-converter design for direct drive wind turbines. In: Proc. IEEE ECCE. 2016

[17] Blaabjerg F, Liserre M, Ma K. Power electronics for wind turbine systems. IEEE Transactions on Industry Applications. 2012;**48**(2):708-719

[18] Lumbreras C, Guerrero JM, Garcia P, Briz F, Reigosa DD. Control of a small wind turbine in the high wind speed

region. IEEE Transactions on Power Electronics. 2016;**30**(10):6980-6991

[19] Dai J, Xu DD, Wu B. A novel control scheme for current-source converter-based PMSG wind energy conversion systems. IEEE Transactions on Power Electronics. 2009;**24**(4):963-972

[20] Tenca P, Rockhill AA, Lipo TA, Tricoli P. Current source topology for wind turbines with decreased mains current harmonics, further reducible via functional minimization. IEEE Transactions on Power Electronics. 2008;**23**(3):1143-1155

[21] Lee JS, Lee KB, Blaabjerg F. Open-switch fault detection method of a back-to-back converter using NPC topology for wind turbine systems. IEEE Transactions on Industry Applications. 2015;**51**(1):325-335

[22] Blaabjerg F, Ma K. Future on power electronics for wind turbine systems. IEEE Journal of Emerging and Selected Topics in Power Electron. 2013;**1**(3):139-151

[23] Zavvos A, Mcdonald A, Mueller M. Optimisation tools for large permanent magnet generators for direct drive wind turbines. IET Renewable Power Generation. 2013;7(2):163-171

[24] Zhang Z, Zhao Y, Qiao W, Qu L. A space-vector-modulated sensorless direct-torque control for direct-drive PMSG wind turbines. IEEE Transactions on Industry Applications. 2014;**50**(4):2331-2441

[25] Potgieter JHJ, Kamper MJ. Design optimization of directly grid-connected PM machines for wind energy applications. IEEE Transactions on Industry Applications. 2015;**51**(4):2949-2958

[26] Singh A, Mirafzal B. Three-phase single-stage boost inverter for direct drive wind turbines. In: 2016 IEEE Energy Conversion Congress and Exposition (ECCE). Milwaukee, WI; 2016. pp. 1-7

[27] Singh A, Mirafzal B. A low-voltage generator-converter topology for direct drive wind turbines. In: 2016 IEEE 7th International Symposium on Power Electron. For Distributed Gener. Sys. (PEDG). Vancouver, BC; 2016. pp. 1-6

[28] Wang Z, Wu B, Xu D, Zargari NR. A current-source-converter-based high-power high-speed PMSM drive with 420-Hz switching frequency. IEEE Transactions on Industrial Electronics. 2012;**59**(7):2970-2981

[29] Mirafzal B, Saghaleini M, Kaviani A. An SVPWM-based switching pattern for stand-alone and grid-connected three-phase single-stage boost inverters. IEEE Transactions on Power Electronics. 2011;**26**(4):1102-1111

[30] Singh A, Milani AA, Mirafzal B. Modified phasor pulse width modulation method for three-phase single-stage boost inverter. In: 2014 IEEE Applied Power Electron. Conf. And Expo. - APEC 2014. Fort Worth, TX; 2014. pp. 1276-1280

[31] Singh A, Kaviani AK, Mirafzal B. On dynamics models and stability of three-phase phasor PWM based CSI for stand-alone applications. IEEE Transactions on Industrial Electronics. May 2015;**62**(5):2698-2707

[32] Holmes DG, Lipo TA. Pulse Width Modulation for Power Converters: Principles and Practice. First ed. New York, USA: John Wiley & Sons; 2003

[33] Fitzgerald AE, Kingsley C Jr, Umans SD. Electric Machinery. Sixth ed. New York, USA: McGraw Hill; 2003

[34] Northern Power NW 1500 Direct-Drive Generator –Report, [Online] Available from: http://www.nrel.gov/docs/fy08osti/40177.pdf

[35] Singh A, Benzaquen J, Mirafzal B. Current source generator–converter topology for direct-drive wind turbines. IEEE Transactions on Industry Applications. March-April 2018;**54**(2):1663-1670

[36] Fateh F, White WN, Gruenbacher D. A maximum power tracking technique for grid-connected DFIG-based wind turbines. IEEE Journal of Emerging and Selected Topics in Power Electronics. 2015;**3**(4):957-966

[37] Yuan X, Li Y. Control of variable pitch and variable speed direct-drive wind turbines in weak grid systems with active power balance. IET Renewable Power Generation. 2014;**8**(2):119-131

[38] IEEE standard for interconnection and interoperability of distributed energy resources with associated electric power systems interfaces. In: IEEE Std 1547-2018 (Revision of IEEE Std 1547-2003). 2018. pp. 1-138

[39] Xiang D, Turu JC, Muratel SM, Wang T. On-site LVRT testing method for full-power converter wind turbines. IEEE Transactions on Sustainable Energy. 2017;**8**(1):395-403

Section 3

Wear and Abrasion of Turbo Machines

Chapter 5

Methodology for Abrasive Wear Evaluation in Elevator Stage Centrifugal Pump Impellers

Rodrigo O.P. Serrano, José G. do V. Moreira, Ana L.P. de Castro, Maria A. Pinto, Edna M. de F. Viana and Carlos B. Martinez

Abstract

The abrasion resistance of the impellers depends on the characteristics of the materials used in their manufacture. In this work, a methodology is proposed for the evaluation of the abrasive wear of the plates of the centrifugal pump impellers, used in the gross water infrastructure station (GWIS) of sedimentary rivers, due to the sediment load variation and the river fluviometric dimension. In order to determine the wear mode and the relationship of the material-specific wear coefficient (K), due to the sediment concentration, a rotating ball abrasometer test was performed on SAE 8620, gray cast iron (GCI), and nodular cast iron (NCI), used in the manufacture of impellers. As an abrasive suspension, the concentrations of 1, 2, 3, 5, and 10 g L^{-1} of sediment were used in distilled water. The wear volume as a function of the relative velocity of the mixture in relation to the impeller blades was estimated mathematically. The results show that: i) The abrasive capacity of the sediments in different concentrations; ii) SAE 8620 steel was more resistant to abrasive wear; and iii) the rotational control of the pump as a function of sediment concentration and river level showed the possibility of reducing wear by 30%.

Keywords: abrasive wear, microabrasion, sediment concentration, hydroabrasion, pumping system

1. Introduction

Gross water elevation stations (GWES), installed in rivers of sedimentary basins, suffer from impeller wear due to the abrasive drag of the sediment load that descends through rivers throughout the year. To understand this process, we analyzed the abrasive wear caused by the sediments in three metal alloys used in the manufacture of centrifugal pump impellers for QWES.

In this study, the sediment concentration variation from 38 to 840 mg L^{-1} of Acre River was considered [1]. In recent work, in which the particle size of the particles pumped by GWES was analyzed, a higher concentration of particles was found between 50 and 100 μm [2].

IntechOpen

The abrasive effects of the sediments may cause furrows on the surfaces of the impellers due to the impact of the particles, increasing the distortion of the grooves due to the increase of the particle diameter [3].

In view of these issues, it was necessary to study the regions of greatest vulnerability to wear, as well as to analyze the evolution of the wear, due to the variation of the sediment concentration at the time, the speed of rotation, variation of the head (ΔH) and of the material used in the manufacture of the rotors.

2. Abrasive wear mechanism

The mechanisms of abrasive wear are part of the tribological studies that study the interactions of surfaces in relative movement [4]. Such interaction takes into account friction, wear, and lubrication, in view of the economic, scientific, and technological interest in minimizing wear and tear (**Figure 1**).

The relationship of friction and wear will depend on the properties and shape of the surface of the material, as well as on the characteristics of the abrasive agent such as the size, shape, and hardness of the particles [5].

According to ASTM G40 [6], wear is the progressive loss of matter from the surface of a solid body due to contact and relative movement with another solid, liquid, or gaseous body. The standard DIN 50320 [7] presents a similar definition and highlights four types of wear mechanisms, among which the abrasive wear (**Figure 2**).

The abrasive wear volume can be calculated by the equation given by Archard and Hirst [8], which is directly proportional to the normal load and the relative

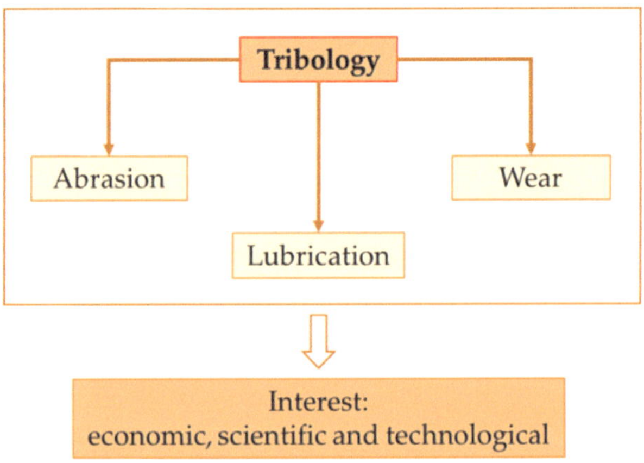

Figure 1.
Flowchart of the interaction of abrasion, wear, and lubrication in the tribological study.

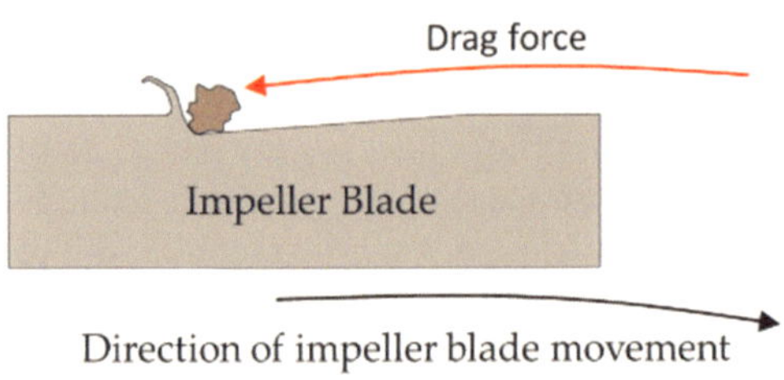

Figure 2.
Mechanism of abrasive wear on impeller blades of centrifugal pumps.

movement distance and inversely proportional to the hardness of the material (Eq. 1).

$$Vw = \frac{k}{\text{Hm}} \bullet FN \bullet S \tag{1}$$

Wear can also be determined by trials in tribosystems, where parameters such as surface properties, chemical composition of the material, slip distance, velocity, and applied load are considered for a better understanding of the wear mechanisms [9, 10].

In pumping systems, wear is the result of the friction that occurs between the continuous water stream and the surface of the impeller blades, in a purely mechanical process [11], classified as hydroabrasive wear.

On the surface of the impeller blades, the microcut can occur due to the multiple impacts of the abrasive particles, being evident that the surface wear intensity depends mainly on the kinetic energy of the particles carried by the flow [12].

3. Methodology

Considering that there are several manufacturers and models of pumps that can work in GWES and the difficulties of acquiring the projects of the appropriate manufacturers, it was decided to carry out this study based on a project impeller. In order to do so, the algorithms developed by Palomino [13], which correlate the main design characteristics of a centrifugal pump impeller, as a function of flow rate (Q), manometric height (H), and rotation of the motor pump system, that is, as a function of the specific rotation (nq), given by Eq. (2).

$$nq = 3.65\frac{n\sqrt{Q}}{H^{3/4}} \tag{2}$$

This algorithm takes into account the sequence proposed in Chapters 4–6, 8, and 10–12 of the book by Macintyre [14], allowing for a quick determination of an impeller designs (Eq. 3–29).

$$\text{Estimate of hydraulic yield } (\varepsilon), \varepsilon = 1 - \frac{0.8}{\sqrt[4]{Q_{gpm}}} \tag{3}$$

$$\text{Driving power } (N_{cv}), N_{cv} = \frac{1000\ QH}{75\eta_t} \tag{4}$$

$$\text{Shaft diameter } (dei), dei = 12\sqrt[3]{\frac{N_{cv}}{n}} \tag{5}$$

$$\text{Shaft diameter—corrected } (dei_{corr}), dei_{corr} = 1.15(dei) \tag{6}$$

$$\text{Core diameter } (dn), dn = dei_{corr} + (2\ (5 \text{ to } 15\ mm)) \tag{7}$$

$$\text{Average speed at the impeller inlet } (v'_1), \nu'_1 = k\nu'_1\sqrt{2gH} \tag{8}$$

$$\text{Diameter of the impeller inlet port } (d'_1), d'_1 = \sqrt{\frac{4Q'}{\pi\nu'_1} + (dn)^2} \tag{9}$$

$$\text{Southern speed at impeller inlet } (v_{m1}), v_{m1} = k_{vm1}\sqrt{2gH} \tag{10}$$

$$\text{Peripheral speed at the leading edge } (u_1), u_1 = \frac{\pi d_{m1} n}{60} \tag{11}$$

$$\text{Shovel inclination angle } \beta_1 \text{ at the impeller inlet, } \tan\beta_1 = \frac{v_{m1}}{u_1} \tag{12}$$

$$\text{Number of blades } (Z), Z = 6,5\left(\frac{d_2 + d_{m1}}{d_2 - d_{m1}}\right) sen\left(\frac{\beta_1 + \beta_2}{2}\right) \tag{13}$$

$$\text{Circumferential pitch between the blades at the impeller inlet } (t_1), t_1 = \frac{\pi d_{m1}}{Z} \tag{14}$$

$$\text{Obstruction due to the blade thickness at the impeller inlet } (\sigma), \sigma_1 = \frac{S_1}{Sen\beta_1} \tag{15}$$

$$\text{Coefficient of contraction } (\Upsilon_1), \frac{1}{\gamma_1} = \frac{t_1}{t_1 - \sigma_1} \tag{16}$$

$$\text{Width at the blade entrance } (b_1)\text{, considering } \sigma 1, b_1 = \frac{Q'}{v_{m1}(\pi d_{m1} - Z\sigma_1)} \tag{17}$$

$$\text{Peripheral speed at output } (u_2), u_2 = k_{u2}\sqrt{2gH} \tag{18}$$

$$\text{Impeller output diameter } (d_2), d_2 = \frac{60\ u_2}{\pi n} \tag{19}$$

$$\text{Desired height (He) for } \varepsilon = 0.87, He = \frac{H}{\varepsilon} \tag{20}$$

$$\text{Lifting height (H'e), adopting : } \psi \text{ from 1.1 to 1.2, } H'e = He\left(1 + \frac{8}{3} \times \frac{\psi}{Z}\right) \tag{21}$$

$$\text{Southern speed at the impeller outlet } (v_{m2}), v_{m2} = k_{vm2}\sqrt{2gH} \tag{22}$$

$$\text{Peripheral speed at impeller outlet } (u_{2(\mathrm{corr})}), u_{2(corr)} = \frac{v_{m2}}{2\ \tan\beta_2} + \sqrt{\left(\frac{v_{m2}}{2}\frac{1}{\tan\beta_2}\right)^2 + gH'_2} \tag{23}$$

$$\text{Reed output diameter of impeller } (d_{2(ret)}), d_{2(ret)} = \frac{60\ u_{2\ (corr)}}{\pi n} \tag{24}$$

$$\text{Circumferential pitch between blades } (t_2), t_2 = \frac{\pi d_{2(ret)}}{Z} \tag{25}$$

$$\text{Impeller blade thickness } (S_1 = S_2) \begin{array}{c} 3 \text{ to } 4 \text{ mm se} : d_2 < 30 \text{ cm; ou;} \\ 5 \text{ to } 7 \text{ mm se} : 30\text{cm} > d_2 > 50\text{cm} \end{array} \tag{26}$$

$$\text{Obstruction due to blade thickness at outlet } (\sigma_2), \sigma_2 = \frac{S_2}{Sen\beta_2} \tag{27}$$

$$\text{Coefficient of contraction } (\Upsilon_2), \gamma_2 = \frac{t_2 - \sigma_2}{t_2} \tag{28}$$

$$\text{Rectified blade width } (b_{2(ret)}), b_{2(ret)} = \frac{Q'}{\pi d_{2(ret)} v_{m2} \gamma_2} \tag{29}$$

To better demonstrate, it is important to understand the velocity triangles considered in the pumping process [13, 14]. Carvalho [15] reports that knowing these speeds is critical to any flow machine design (hydraulic turbines, gas turbines, fans, rotary compressors, hydraulic pumps, etc.).

Finally, the algorithm provides the constructive characteristics that were used to calculate the forces acting on the impeller blades. Based on the rotation variation of the impeller of 100, 95, 90, 85, and 80%, maintaining the specific rotation (nq) and flow (Q) constant, it was possible to calculate the variation of the rotation of the impeller [9].

3.1 Determination of forces on impeller blades

Based on the constructive characteristics of the impeller, the length of the blade was determined, which, in turn, was divided into five sections to determine the acting forces (**Figure 3**).

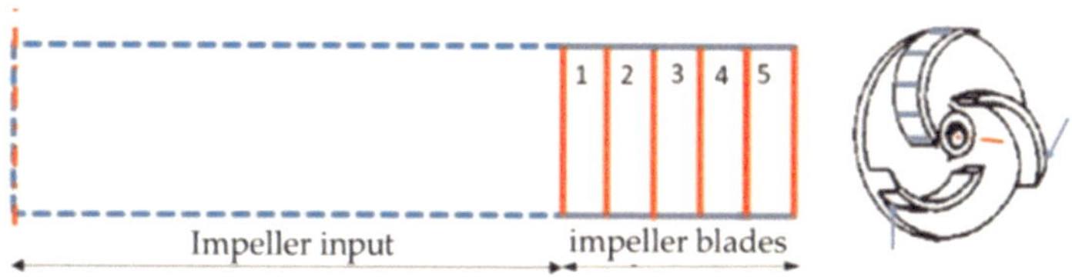

Figure 3.
Diagram of division of impeller blade sections.

Based on the velocity fields that occur in each section of the impeller blade, the abrasive drag forces occurring on the blade surface were determined. According to Fox [16], drag is the component of force on a body that acts parallel to the direction of relative motion and can be written as presented in Eq. (30).

$$F_D = \frac{Cd\rho V^2 A}{2} \tag{30}$$

Considering that the flow is initially laminar and in the course of the flow area, it becomes turbulent, the drag coefficient can be calculated by Eq. (31).

$$C_D = \frac{0.0742}{\text{Re}^{(0.2)}} - \frac{1740}{\text{Re}} \tag{31}$$

Fox [16] also reports that in the case of "$5 \times 10^5 < \text{Re} < 10^9$", the same adjustment performed in Eq. (31) can be applied to Eq. (32) given by Schlichting in 1979, obtaining Eq. (33):

$$C_D = \frac{0.455}{(\log \text{Re})^{2.58}} \tag{32}$$

Staying like this:

$$C_D = \frac{0.455}{(\log \text{Re})^{2.58}} - \frac{1740}{\text{Re}} \tag{33}$$

The Reynolds number (Re) is a relation between forces of inertia and viscous force, which can be expressed according to Eq. (34):

$$\text{Re} = \frac{dV}{\nu} \tag{34}$$

3.2 Tests of resistance to abrasive wear

In this test, the sample was pressed by loading an inoperative weight against a rotating steel ball. The abrasive slurry was dripped onto the wear interface. After the sphere travels through a previously established distance, called a running distance, a circular depression or crater of wear occurs on the surface of the sample [9, 10, 17–21]. **Figure 4** shows a schematic diagram of the microabrasion test device.

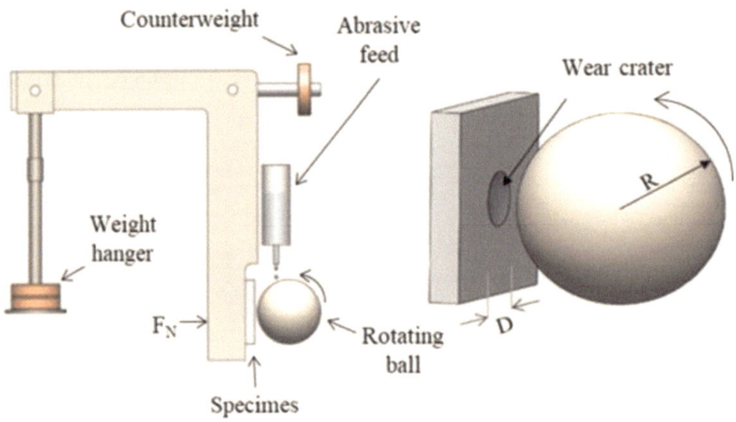

Figure 4.
Schematic diagram of the microabrasion test device.

The equations for calculating the wear coefficient from the volume of material removed in tests with spherical body were initially established by Kassman et al. [22]. Rutherford and Hutchings [23] generalized these equations for the case of flat and curved samples and for calculating the coefficient of wear of films independently of the substrate.

After the wear tests, the simplest technique for the calculation of the wear coefficient is based on the reading of the average diameter of the crater formed by the wear followed by the application of Eqs. (35)–(37) presented below [9, 10, 17, 18, 20, 21]. The diameter measurement should be performed by scanning electron microscope (SEM).

$$\text{Theoretical volume removed } (V_{TW}),\ V_{TW} = \frac{\pi D^4}{64.R} \quad (35)$$

$$\text{Wear rate } (Q_T),\ Q_T = \frac{V_{TW}}{S} \quad (36)$$

$$\text{Coefficient of material wear } (K),\ K = \frac{Q_T}{FN} \quad (37)$$

Studies suggest the application of the following test parameters: 25.4 mm diameter test sphere made of SAE 52100 steel; rotation of 100 rpm; duration of 23 minutes for each trial; scroll distance of 180 m; and FN = 3 N [9, 10].

3.3 Acquisition and preparation of analyzed samples

Of the three types of materials used in impeller manufacture (**Figure 5**), only nodular cast iron (NCI) was removed from a genuine impeller of a Flygt pump. The other samples were acquired in bar format with specifications similar to those used in the manufacture of impellers.

The samples were sanded, polished, and cleaned. For this work, the technique of manual sanding was chosen, where the samples were sanded with granulometry ranging from 100# to 1200#, changing the direction by 90° in each change of

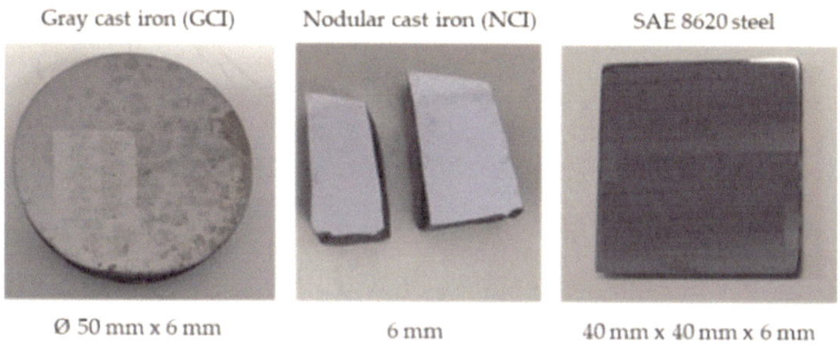

Figure 5.
Samples analyzed in the abrasion tests.

sandpaper, being careful to change the sandpaper only when the traces of the previous sandpaper did not appear (**Figure 6**) [24].

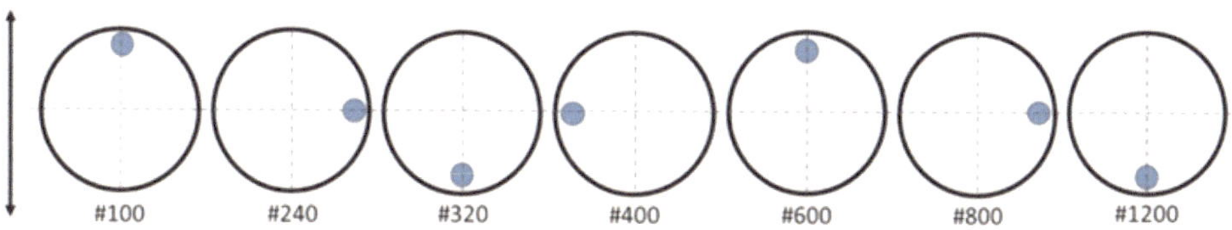

Figure 6.
Schematic representation of the sanding method.

After sanding, the samples were subjected to a 1-μm alumina polishing procedure at a concentration of 10%, followed by polishing with a 1-μm diamond paste and 0.25 μm for a nonmarking surface finish (**Figure 7**).

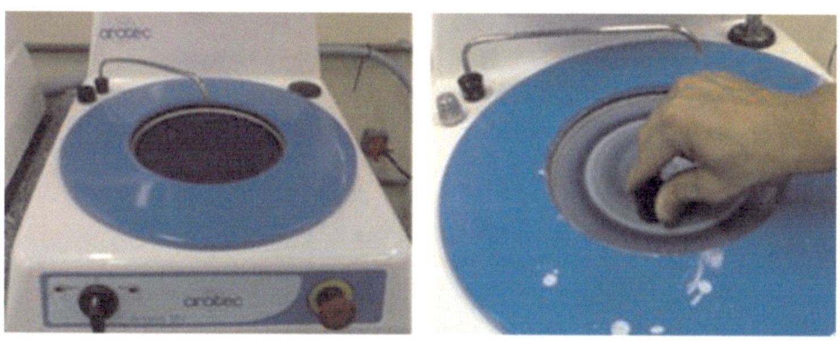

Figure 7.
Polishing the sample in bench polisher.

The samples were then washed in tap water to remove traces of the abrasives, followed by the application of ethyl alcohol to facilitate drying.

3.4 Preparation of abrasives

Each sediment sample was shaken with distilled water for homogenization in five different concentrations 0.1, 0.2, 0.3, 0.5, and 1.0% (1, 2, 3, 5, and 10 g L^{-1}, respectively) (**Figure 8**).

To compare the abrasiveness of the sediments and the abrasive wear resistance of test bodies, a 1% concentration (10 g L^{-1}) abrasive suspension of silicon carbide (SiC) was also prepared (**Figure 9**). This procedure allows to record a reference abrasiveness caused by the sediment used in comparison of a standard abrasive, commonly used in other tests [9, 10].

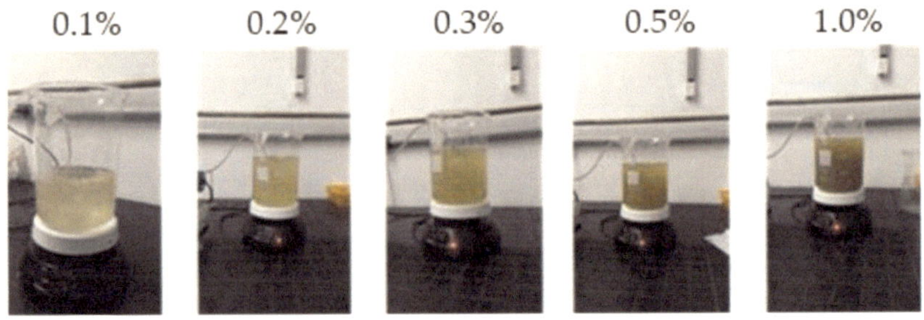

Figure 8.
Preparation of abrasive suspensions with Acre River sediments.

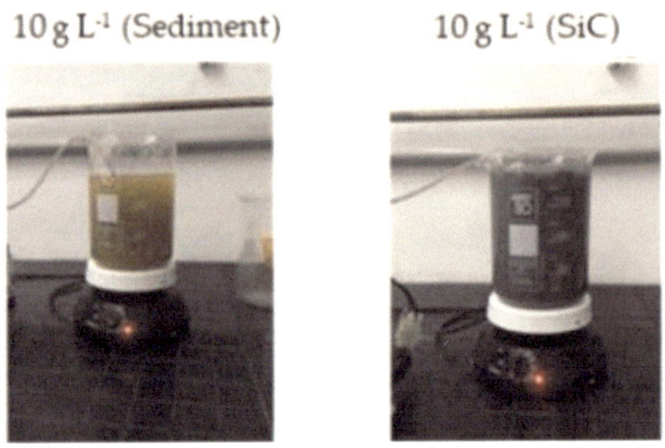

Figure 9.
Preparation of abrasive solutions (sediment and SiC).

To obtain the granulometric distribution of sediment particles, the combined analysis of the sedimentation and sieving method was used [25].

3.5 Impulse wear estimate

The estimated wear of the impeller blades was predicted based on the modified Archard equation [20], where the coefficient of wear (k) presented in Eq. (1) is a dimensionless constant that is divided by the hardness of the worn material (H) and that the K displayed in Eq. (37) refers to a specific wear coefficient for a material as a function of an abrasive, allowing the following adjustment (Eq. 38) [9, 10]:

$$V_{TW} = K \bullet FN \bullet S \tag{38}$$

Considering the relative velocity of the mixture (water + sediment) passing through a certain area of the impeller blade, a relative distance (S) was estimated according to Eq. (39) [9, 10]. The FN force considered in this analysis was calculated from Eq. (30) of the drag force occurring in a particular section of the blade.

$$S = V \bullet t \tag{39}$$

Based on the variation of the Acre River account [26], which changes manometer height (H) of the GWES, wear was determined as a function of the variation of the pump rotation in 100, 95, 90, 85, and 80%, due to the variation of the head (ΔH), maintaining the volume of water pumped. The reduction of the FD due to the reduction of the relative speed (w), due to the reduction of the rotation of the pump, to estimate the variation of the wear of the impeller blades due to the variation of the sediment load and the Acre river quota was observed.

4. Results and discussion

4.1 Impeller characteristics

To determine the characteristics of the impeller, consideration was given to the need of GWES of the water treatment station II (WTS II), which works with pumps with flow of 1080 $m^3 h^{-1}$ (300 $L s^{-1}$) driven by an electric motor with rotation of 1180 rpm and a head of 25 m (considering hydraulic losses). The results can be seen in **Table 1** and **Figure 10**.

Based on **Table 1**, the propeller length of 140.2 and 7 mm thickness was determined (**Figure 10A** and **B**). The impeller blade has been divided into five sections to determine the abrasive drag forces (**Figure 10C**), which vary from inlet to outlet inlet.

Then, five conditions were simulated, where the flow (Q), the specific rotation (nq), was maintained to relate the rotation (n) of the pump (80, 85, 90, 95 and 100% of 1180 rpm), as a function of the variation of the head (ΔH) Acre (**Figure 11**).

Impeller characteristics	
Diameter of impeller inlet (d1)	312 mm
Impeller outlet diameter (d2)	408 mm
Angle of blade at impeller inlet (β1)	17.45°
Angle of blade at impeller outlet (β2)	23°
Blade width at impeller inlet (b1)	99 mm
Width of blade at impeller outlet (b2)	63 mm
Thickness of impeller blades	6 mm
Number of impeller blades	7
Southern speed at impeller inlet (vm1)	4.89 $m s^{-1}$
Southern speed at impeller outlet (vm2)	4.27 $m s^{-1}$
Peripheral velocity at impeller inlet (u1)	15.57 $m s^{-1}$
Peripheral speed at impeller outlet (u2)	25.18 $m s^{-1}$
Relative speed at impeller inlet (w1)	16.31 $m s^{-1}$
Relative speed at impeller outlet (w2)	10.94 $m s^{-1}$

Table 1.
Constructive characteristics calculated.

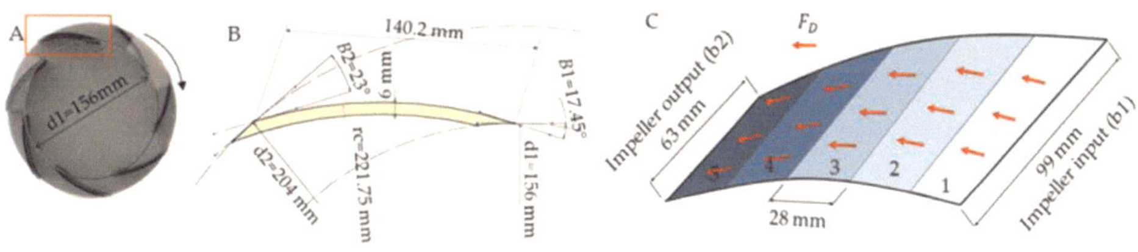

Figure 10.
Determination of the five sections of the impeller blade to determine the abrasive drag forces on the blade surface.

4.2 Determination of forces applied to impeller blades

As shown in **Figure 10** and applying Eq. (30) and the variation of the relative speed (w) shown in **Figure 11**, the abrasive drag force (FD), which varies as a function of the impeller radius in each section of the sheet, is calculated. In this way, it can be observed that the FD is larger in Section 2, with a tendency to decrease toward the exit of the impeller blades (**Figure 12**).

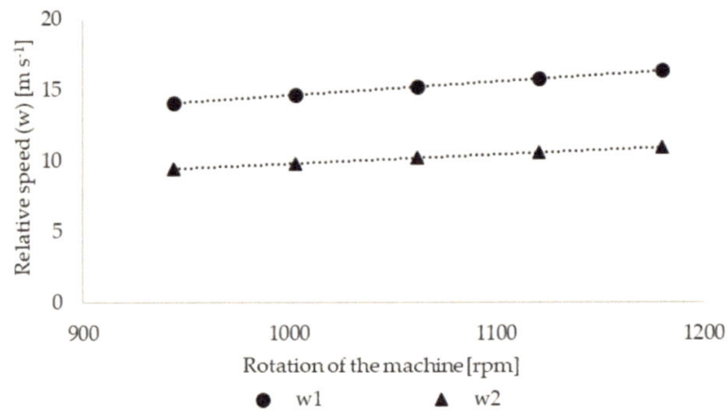

Figure 11.
Variation of the relative speed (w) in the impeller blade, due to the variation of the pump rotation: w1 inlet and w2 outlet.

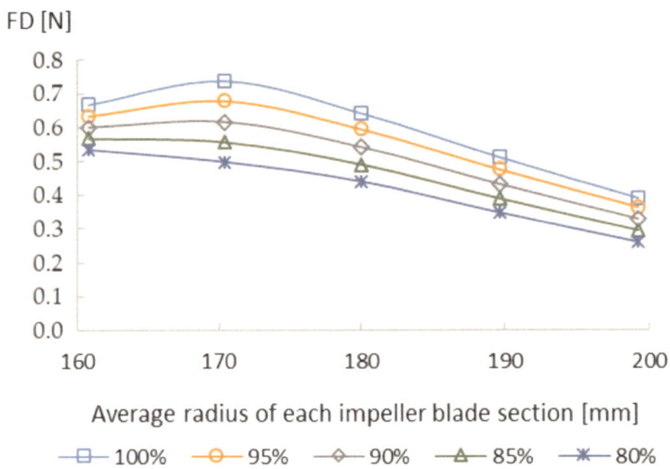

Figure 12.
Variation of the FD to the radius of each section of the impeller blade and the rotation variation from 80 (944 rpm) to 100% (1180 rpm).

4.3 Analysis of the erosive sediments of Acre River

The particles pumped by the impellers may show differences of granulometry and concentration due to the effect of turbulence at the entrance of the uptake ducts of said elevation stations, due to the slight dredging effect. In the case of the samples collected in the desanding tank, after the pumping, they presented a concentration of 60% of the sediment load with granulometry varying from 0.075 to 0.420 mm, with 1.36% having a granulometry greater than 0.420 mm and the remaining 40.35% with particle size below 0.075 mm diameter D50 of 0.097 mm (**Figure 13**).

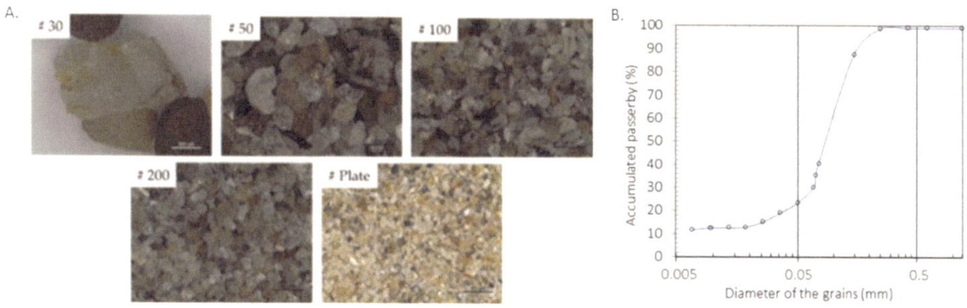

Figure 13.
Classified morphology of the erosive sediments present in the waters of the Acre River, collected from the WTS II degrading tank.

The sediments presented granulometric classification of very fine to fine sand [27] or simply sand [28]. According to visual interpretation, sediment grains are classified as semiround to semiround, characteristic of young and sandy soils. These characteristics indicate a high abrasive power due to the sharp tips and the large quantity of quartz grains with a hardness of 4–6 on the Mohs scale.

4.4 Rotating ball abrasion tests and crater analysis

From the diameters of the craters generated in the wear tests for the sediment concentrations analyzed, the coefficient of wear (K) was calculated from Eq. (37).

To analyze the relationship of K as a function of the concentration of sediment in the abrasive suspension, for each of the analyzed materials, logarithmic curves were drawn (**Figure 14**). The values of K at the concentration of 0.5 g L^{-1} were calculated by the curve equation, generated experimentally for the concentrations of 1, 2, 3, 5, and 10 g L^{-1}. Thus, it was observed that K increased as a function of the increase of the sediment concentration, for FN = 3N, probably due to the higher amount of abrasive particles in contact between the sample and the sphere, resulting in higher wear [9, 10, 21, 29–32].

It is also observed that the increase of K does not present a linear proportion with the increase of the concentration, considering that not all the particles of the abrasive suspension can come into contact with the worn surface, corroborating with that described by Hutchings and Shipway [33].

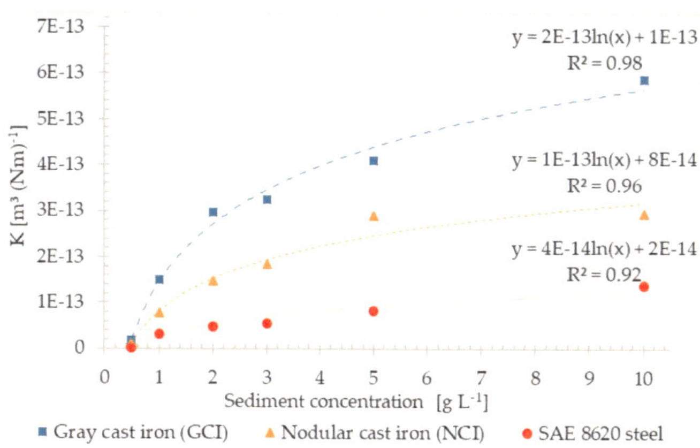

Figure 14.
Behavior of the specific wear coefficient (K) due to the sediment concentration variation, for the test conditions (S = 180 m and FN = 3N).

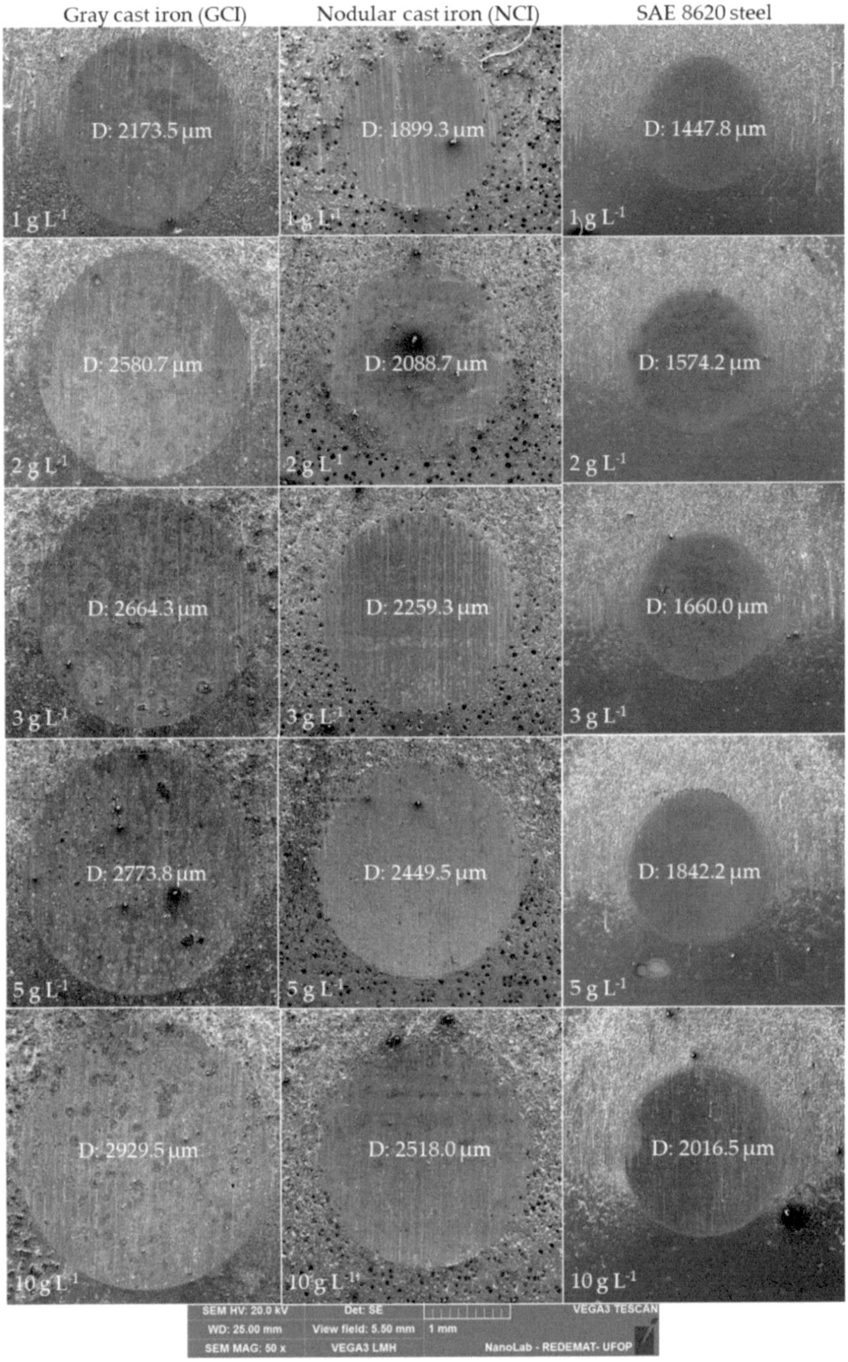

Figure 15.
Influence of sediment concentration on the diameter of the wear cap (SEM).

Figure 15 shows the wear evolution due to the sediment concentration, as well as the difference in crater diameter, for each test specimen used. At the outset, it is possible to observe the presence of wear grooves characteristic of abrasive tests.

Figure 16.
Evaluation of the wear pattern as a function of sediment concentration (SEM): (A) plastic deformation and (B) tipping points of that of the accumulations in front of the grooves.

From what is shown in **Figure 16**, it can be observed that the wear pattern did not change as a function of the sediment concentration, maintaining the wear pattern by grooving for the three metal alloys, with a clear orientation. These results also reflect features similar to the wear pattern found by Cozza [34] and Serrano et al. [9], referring to concentrations of abrasive less than 18% and a load greater than 1N.

In addition, it is observed in **Figures 15** and **16** that the GCI and NCI specimens presented areas with clear plastic deformations (A), with intensification as a function of the increase in sediment concentration, with a clear accumulation of material at the front of the wear grooves and with greater intensity of deformation in the GCI. The materials accumulated at the edges of the grooves are later removed by the association of the microstructural mechanisms with the mechanism of low cycle fatigue (B), similar to that observed by Serrano et al. and Pintaúde [9, 10, 31].

The average volume of waste material caused by the sediment was 340% higher, compared to the wear caused by the SiC. The results can be explained by the granulometric difference of the abrasive materials, where the sediment particles presented D50 six times larger than the SiC particles, 0.097 and 0.016 mm, respectively. Considering the homogeneity of SiC hardness and grain size, it will allow the comparison of the wear intensity caused by sediments from different river basins [9].

4.5 Determination of blade wear

Considering the relative velocities at the inlet and outlet of the impeller (**Figure 11**), the variation of FD on the blades (**Figure 12**) and the variation of K as a function of the sediment load (**Figure 14**) were applied to Eq. (38) to estimate the paddle wear volume, due to the variation of the sediment load and the monthly mean level of the Acre River (**Figure 17**).

As shown in **Figure 17**, the nonapplication of the rotational control of the machine, taking advantage of the reduction of the ΔH as a function of the increase of the river's elevation, can lead to premature wear of the impeller due to the increase of the sediment concentration carried by river, the during the rainy season.

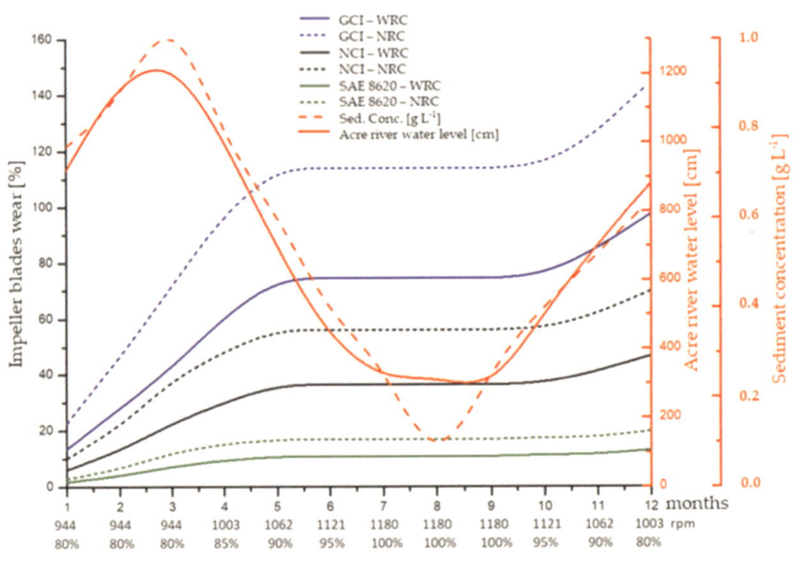

Figure 17.
Accumulated wear of the impeller blade, due to the variation of the sediment concentration of the Acre River: with rotational control (WRC) and without rotational control (NRC).

In the case of the study presented here, when applying rotational control, the relative velocity of the fluid decreased, reducing the FD on the impeller blades, also reducing the wear.

When looking at the wear accumulated in **Figure 17** and considering the wear limit at 70% of the thickness of the impeller blades, it can be concluded that without the rotational control, it would require two GCI impellers per year or one NCI impeller, while the SAE 8620 steel impeller could run for almost 3 years. Already with the rotational control, this wear can be reduced by approximately 33%.

Also in **Figure 17**, the higher resistance of SAE 8620 steel can be verified, being 7.3 times more resistant than GCI and 3.5 times more resistant than NCI, whereas NCI is 2 times more resistant than the GCI.

5. Conclusions

The abrasive wear regime and the reproducibility of the results under the test conditions in the present work were adequate to investigate the abrasive capacity of the sediments and the wear of GCI, NCI, and ASE 8620 steel.

The semiangle and semirounded forms of the Acre River sediments produced evidence of microstrip and plastic deformation of the material in the three metal alloys at all sediment concentrations used in the tests.

The analysis of the wear volume, according to the type of metal alloy, the sediment concentration, and the relative velocity in the impeller blades, showed the importance of a previous study of the sediment concentration in the water sources pumped by GWES, as well as ΔH and the required flow rate, to select the material used to make the impeller.

The analysis of the variation of the impeller rotation as a function of the river level variation and the variation of the head (ΔH) showed the possibility of reducing the impeller wear by up to 30% in a hydrological cycle for the three metal alloys used in the tests, showing the importance of rotational control during the pumping process.

Considering that the three metal alloys presented the same percentages of wear reduction as a function of the rotational control, maintaining the same volume of water pumped, it is suggested that such control can be adopted in GWES from other locations that show load variation of sediment and fluviometric dimension.

Acknowledgements

The authors are grateful for the financial support: Gorceix Foundation; CEMIG; ANEEL; FAPEPIG; ELETROBRAS-FURNAS; CAPES; and CNPq.

They also thank the Laboratory of Thermal Treatment and Optical Microscopy (LTM), the Laboratory of Electron Microscopy (NanoLab), the Laboratory of Microabrasive Foundry and Wear of UFOP, the Center for Hydraulic Research and Water Resources (CPH) of UFMG, and the Laboratory of Geomorphology and Sedimentology of UFAC, for technical and experimental support.

Nomenclature

V_W	volume of material removed (m^3)
k	coefficient of wear (dimensionless)
Hm	hardness of material ($N\ mm^{-2}$)
FN	normal force applied (N)

S	sliding distance (m)
nq	specific rotation (rpm)
n	impeller rotation (rpm)
Q	pump flow rate (L s^{-1})
H	manometric height (m)
Q_{gpm}	pump flow rate (gpm)
gpm	gallons per minute
ηt	absolute yield
kv'1	the coefficient of velocity at the inlet port of the impeller
g	gravitational constant (m s^{-2})
π	PI
Q'	correction pressure: due to recirculation (m^3 s^{-1})
K_{vm1}	coefficient of the velocity of the meridian component at the impeller inlet
F_D	abrasive drag force (N)
C_d	drag coefficient
ρ	fluid density (kg m^{-3})
V	fluid relative velocity (m s^{-1})
A	area (m^2)
d	length of fluid displacement (m)
ν	kinematic viscosity (m^2 s^{-1})
D	wear crater diameter (m)
R	ball radius (m)
Q_T	wear rate (m^3 m^{-1})
K	specific wear coefficient (mm^3 $(N\ m)^{-1}$)
V_{TW}	theoretical volume removed (m^3)
t	impeller run time (s)
ΔH	variation of manometric height (m)
K_{vm2}	coefficient of velocity of the meridian velocity component at the impeller outlet

Author details

Rodrigo O.P. Serrano[1,4*], José G. do V. Moreira[2], Ana L.P. de Castro[3], Maria A. Pinto[3], Edna M. de F. Viana[4] and Carlos B. Martinez[4]

1 Federal University of Acre—UFAC, Rio Branco, Acre, Brazil

2 Federal University of Acre—UFAC, Cruzeiro do Sul, Acre, Brazil

3 Federal University of Ouro Preto—UFOP, Ouro Preto, Minas Gerais, Brazil

4 Federal University of Minas Gerais—UFMG, Program in Mechanical Engineering, Belo Horizonte, Minas Gerais, Brazil

*Address all correspondence to: ropereas@gmail.com

IntechOpen

© 2019 The Author(s). Licensee IntechOpen. This chapter is distributed under the terms of the Creative Commons Attribution License (http://creativecommons.org/licenses/by/3.0), which permits unrestricted use, distribution, and reproduction in any medium, provided the original work is properly cited. (cc) BY

References

[1] Carvalho AT, Costa ML, Almeida HDF. Os Sedimentos em Suspenção dos Rios Purus e Juruá no Estado do Acre. Revista Científica da UFPA, v.7. 2008:8

[2] Serrano ROP, Ferreira-Junior AG, Castro ALP, Santos PABV, Menezes MV, Martinez CB. Desgaste do rotor por abrasão: O efeito do bombeamento de água bruta com diferentes cargas de sedimento. In: XXVII Congreso Latinoamericano de Hidráulica, AIRH. Lima, Perú; 2016. pp. 26-30

[3] Xing D, Hai-lu Z, Xin-yong W. Finite element analysis of wear for centrifugal slurry pump. In: Proceedings of the international conference on mining science and technology (ICMST). Procedia Earth and Planetary Science. 2009;**1**:1532-1538. DOI: 10.1016/j.proeps.2009.09.236

[4] Hutchings IM. Tribology: Friction and Wear of Engineering Materials. 1st ed. Londres: Edward Arnold; 1992. 273 p

[5] Upadhyay RK, Kumaraswamidhas LA. A review on tribology of surfaces and interfaces. Advanced Materials Letters. 2014;**5**(9):486-495. DOI: 10.5185/amlett.2014.5566

[6] ASTM—G40. Standard terminology relating to wear and erosion. In: G 40, Annual Book of ASTM Standards. ASTM International; 2015

[7] DIN. DIN 50320. Wear; terms, systematic analysis of wear processes, classification of wear phenomena. International Classification for Standards. Alemannic; 1979. 8 p

[8] Archard JF, Hirst W. The wear of metals under unlubricated conditions. Proceedings of the Royal Society. 1956; **236**:397-410. DOI: 10.1098/rspa.1956.0144

[9] Serrano ROP, Santos LP, Viana EMF, Pinto MA, Martinez CB. Case study: Effects of sediment concentration on the wear of fluvial water pump impellers on Brazil's Acre River. Wear. 2018;**408–409**:131-137. DOI: 10.1016/j.wear.2018.04.018

[10] Serrano ROP, Castro ALP, Rico EAM, Viana EMF, Pinto MA, Martinez CB. Abrasive effects of sediments on impellers of pumps used for catching raw water. Revista Brasileira de Engenharia Agrícola e Ambiental. 2018; **22**(9):591-596. DOI: 10.1590/1807-1929/agriambi.v22n9p591-596

[11] Duan CG. Approach to the demand on anti abrasive Erosion from hydraulic machinery project. Hydraulic machinery and cavitation. In: Proceedings of XIX IAHR Symposium, World Science. Singapore; 1998. pp. 59-69

[12] Serrano ROP. Metodologia para avaliação de desgaste abrasivo em pas de rotor de bombas centrifugas de estação elevatória [thesis]. Belo Horizonte, Minas Gerais, Brazil: Federal University of Minas Gerais; 2017

[13] Palomino AEC. Desenvolvimento de metodologia para determinação dimensional de uma bomba centrífuga utilizando velocidades específica [dissertation]. Bolo Horizonte, Minas Gerais, Brazil: Universidade Federal de Minas Gerais; 2017

[14] Macintyre AJ. Bombas e instalações de bombeamento. 2nd ed. Rio de Janeiro: LTC; 2013. 430 p

[15] Carvalho DF. Instalações elevatórias – Bombas. FUMARC – Fundação Mariana Resende Costa; 2010. 365 p

[16] Fox RW, McDonald AT, Pritchard PJ. Introdução à mecânica de fluidos. Tradução de Ricardo Nicolau Nassar Koury, Geraldo Augusto Campolina.

Vol. 2006. Rio de Janeiro: França, LTC; 1934. 513 p

[17] Rutherford KL, Hutchings I, Rutherford K. Theory and application of a micro-scale abrasive wear test. Journal of Testing Evaluation of the American Society for Testing and Materials. 1997;**25**(2):250-260. DOI: 10.1520/JTE11487J

[18] Allsopp DN, Hutichings IM. Micro-scale abrasion and scratch response of PVD coatings at elevated temperatures. Wear. 2001;**251**:1308-1314. DOI: 10.1016/S0043-1648(01)00755-4

[19] Cozza RC, Rodrigues LC, Schon CG. Analysis of the micro-abrasive wear behavior of an iron aluminide alloy under ambient and high-temperature conditions. Wear. 2015;**330-331**: 250-260. DOI: 10.1016/j.wear.2015.02.021

[20] Santos WC, Pereira-Neto JO, Silva RO, Rodrigues G, Moreto JÁ, Manfrinato MD, et al. Desenvolvimento de dispositivo e estudo do comportamento ao micro desgaste abrasivo do aço AISI 420 temperado e revenido. Revista Matéria. 2015;**22**(02): 304-315. DOI: 10.1590/S1517-707620150002.0031

[21] Krelling AP, Costa CE, Milan JCG, Almeida EAS. Micro-abrasive wear mechanisms of borided AISI 1020 steel. Tribology International. 2017;**111**: 234-242. DOI: 10.1016/j.triboint.2017.03.017

[22] Kassman A, Jacobson S, Erickson L, Hedenqvist P, Olsson M. A new test method for the intrinsic abrasion resistance of thin coatings. Surface and Coatings Technology. 1991;**50**:75-84. DOI: 10.1016/0257-8972(91)90196-4

[23] Rutherford KL, Hutchings IM. A micro-abrasive wear test with particular application to coated systems. Surface and Coatings Technology. 1996;**79**: 231-239. DOI: 10.1016/0257-8972(95)02461-1

[24] Rohde RA. Metalógrafia—Preparação das amostras: Uma abordagem pratica. LEMM: URI; 2010. 30 p

[25] ABNT—Associação Brasileira de Normas Técnicas. NBR 7181—Solo—Análise granulométrica—Métodos de ensaios. 1984

[26] Duarte AF, Gioda A. Inorganic compositional of suspended sediments in the Acre River, Amazon Basin, Brazil. Latin American Journal of Sedimentology and Basin Analysis. 2014;**21**(1):3-15

[27] Wentworth CK. A scale of grade and class terms for clastic sediments. Journal of Geology. 1922;**30**:377-392

[28] ABNT—Associação Brasileira de Normas Técnicas. NBR 6502—Rochas e solos. Brazil; 1995. 18 p

[29] Cozza RC. Estudo do desgaste e atrito em ensaios micro-abrasivos por esfera rotativa fixa em condições de força normal constante e pressão constante [thesis]. Escola Politécnica da USP; 2011

[30] Trezona RI, Allsopp DN, Hutchings IM. Transitions between two-body and three-body abrasive wear: Influence of test conditions in the microscale abrasive wear test. Wear. 1999;**225-229**: 205-214. DOI: 10.1016/S0043-1648(98)00358-5

[31] Pintaúde G. Análise dos regimes moderado e severo de desgaste abrasivo utilizando ensaios instrumentados de dureza [thesis]. Escola Politécnica da USP; 2002

[32] Fernandes F, Ramalho A, Loureiro A, Cavaleiro A. Mapping the micro-abrasion resistance of a Ni-based coating deposited by PTA on gray cast iron.

Wear. 2012;**298**(293):151-158. DOI: 10.1016/j.wear.2012.05.018

[33] Hutching IM, Shipway P. Tripology: Friction and Wear of Engineering Materials. 2nd ed. Butterworth Heinemann; 2017. 412 p. e-book ISBN: 9780081009512

[34] Cozza RC. Effect of pressure on abrasive wear mode transitions in micro-abrasive wear tests of WC-Co P20. Tribology International. 2013;**57**: 266-271. DOI: 10.1016/j.triboint. 2012.06.028

Section 4

Vertical Axis Wind Turbines

Chapter 6

Straight-Bladed Vertical Axis Wind Turbines: History, Performance, and Applications

Yan Li

Abstract

Wind turbine is a kind of rotating machinery. Although the horizontal axis wind turbine (HAWT) is the most popular wind turbine, the vertical axis wind turbine (VAWT) with the main advantages of smart design, novel structure, and wind direction independence receives more and more attention in small-scale wind power market. The straight-bladed VAWT (SB-VAWT) is one of the most researched and studied VAWTs. In this chapter, the historical development of the SB-VAWT will be briefly reviewed firstly. Then the aerodynamic models for the turbine design and performance analysis will be introduced. Finally, the types of traditional and new SB-VAWT and their characteristics and main utilizations will be introduced.

Keywords: vertical axis wind turbine, straight-bladed, aerodynamics performance, wind tunnel test, numerical simulation, application

1. Introduction

Wind turbine is a kind of rotating machine which can convert wind energy to mechanical energy, thermal energy, or electrical energy. With the years of research and development, many kinds of wind turbine were proposed and invented. There are some ways to classify different kinds of wind turbines. Generally, wind turbines can be classified by application, capacity, blade number, relative position of rotor shaft to ground, type of aerodynamic forces generated by blade, and so on. Among them, the relative position of rotor shaft to the ground and the aerodynamic forces of blade are the two main ways for classification of wind turbine. According to the relative position of rotor to ground, wind turbines can be classified into horizontal axis wind turbine (HAWT) and vertical axis wind turbine (VAWT). According to the force type of the blade, wind turbines can be classified into lift-type wind turbines and drag-type wind turbines. The main kinds of wind turbines are shown in **Figure 1** [1].

Based on the progress of aerodynamic theory of aircraft, the propeller-type HAWT becomes the most popular lift-type wind turbine in the world. Nevertheless, the VAWT has not been paid so much attention in the past. In this modern time, benefiting from the rapid progress of large-scale wind turbine and wind farm, the small-scale wind turbine which can be used for distributed generation and off-grid wind power market has also received more and more attention. However, there are some disadvantages of small-scale wind turbine comparing with the large-scale

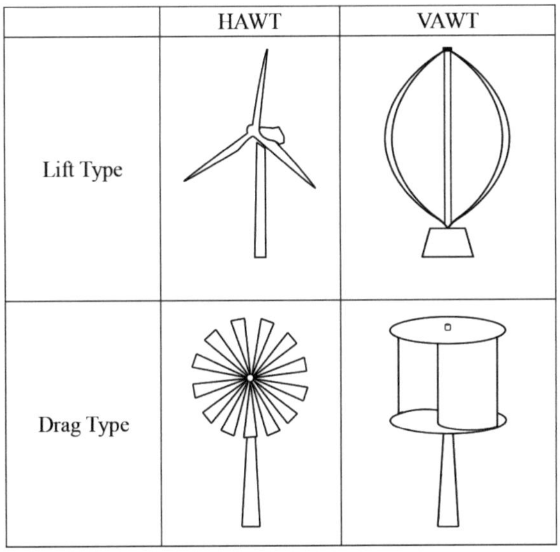

Figure 1.
Main categories of wind turbine [1].

one, such as small swept area of rotor, difficulty of starting at low wind speed, and getting affected easily by wind speed and direction. For HAWT, the rotor should always face the wind direction. Therefore, a yaw system must be equipped. However, the VAWT can receive wind from any direction and does not need a yaw system. This is the biggest difference between the HAWT and VAWT. Due to its characteristics of independence from wind direction comparing with the HAWT, the VAWT can be thought as a good choice for the small- and middle-scale wind power market [2].

Although there are some types of VAWT, the straight-bladed vertical axis wind turbine (SB-VAWT) as a kind of lift-type VAWT with the main advantages of simple design, low cost, and good efficiency becomes one of the most researched and studied VAWTs. The SB-VAWT can be thought as a good choice for serving to the off-grid small-scale energy system which is one of the most important renewable energy resources for the people in urban area and countryside [2].

In this chapter, the historical development of the SB-VAWT will be briefly reviewed firstly. Then the aerodynamic models for the turbine design and performance analysis will be introduced. The research methods including wind tunnel test, visualization test, and numerical simulation are also introduced. Finally, the main utilizations and applications are introduced briefly.

2. Brief history of the development of SB-VAWT

2.1 The development of VAWT

Wind energy has been used for centuries for mankind. The earliest practical wind machines which could be found in documentary records were the vertical axis windmills [1]. These windmills were almost drag type with a multibladed rotor operating at very low tip-speed ratios (much less than unity), which explains their poor efficiency. In spite of the simple design, these types of wind turbines need large amounts of material, are not able to withstand high wind loads, and thus have not proven cost-effective. At the end of the nineteenth century, there were more than 30,000 wind mills in Europe, used primarily for the milling of grain and water pumping [3].

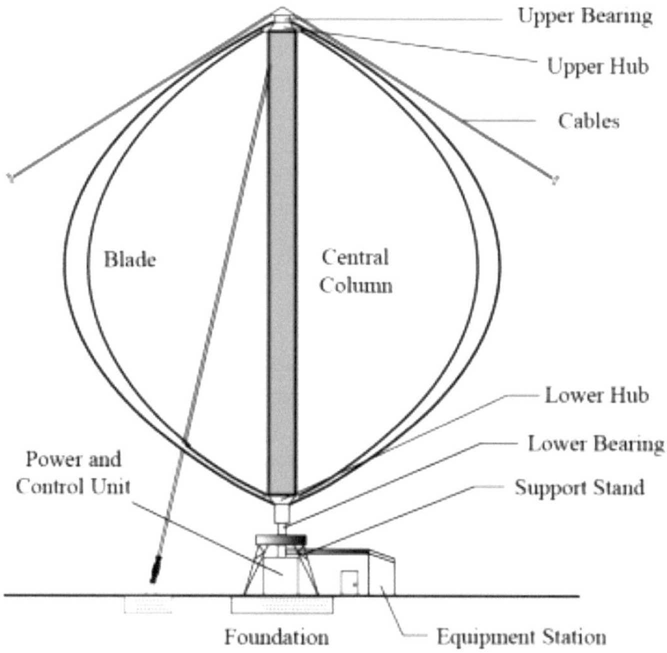

Figure 2.
Darrieus-type VAWT [3].

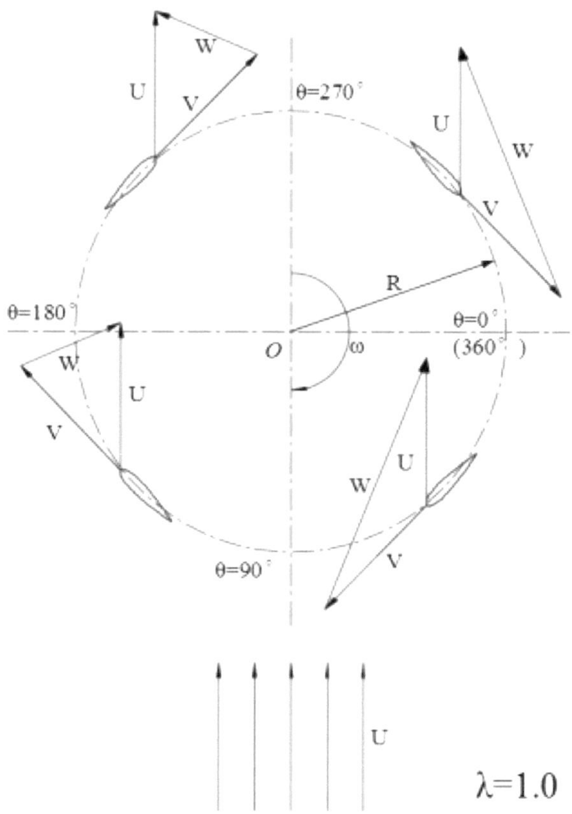

Figure 3.
Blade local relative velocities at different rotating angles (where U is wind speed, V is rotational speed of blade, and W is relative speed).

There was a golden development period of VAWT in 20 ~ 30s of the 20th century. At that time, some typical VAWTs were proposed and researched, mainly including Savonius rotor, Madaras rotor, Darrieus VAWT, and so on. The Darrieus VAWT is the most important one [3].

The Darrieus VAWT was invented in 1931 by a French engineer named George Jeans Marie Darrieus [4]. In the patent he proposed, there were two kinds of rotors included both the "curved blade" and "straight-bladed." Usually, the wind turbine with curved blade is just called as Darrieus VAWT. The turbine rotor with straight blade is called straight-bladed Darrieus-type VAWT, or straight-bladed VAWT simply. The Darrieus VAWT is basically a lift-type wind turbine. The rotor consists of two or more airfoil-shaped blades which are attached to a rotating vertical shaft. **Figure 2** shows a typical Darrieus VAWT according to the book written by Prof Ion Para [3]. The Troposkien curve is often selected for the curved blade shape, which can minimize the bending stress in the blades. The changes of local relative velocity of blade during a revolution are shown in **Figure 3**. Darrieus VAWT was commercially deployed in the USA and Canada in the past.

2.2 The SB-VAWT

The SB-VAWT can be thought as a kind of Darrieus VAWT whose curved blades are replaced by straight blades [3]. Therefore, it always was called as H-Rotor for its rotor outlook. Compared with the Darrieus VAWT, the rotor structure of SB-VAWT is simpler, and the cost of manufacture is cheaper. **Figure 4** shows the structure of a typical SB-VAWT. Generally, a normal SB-VAWT usually has 2–6 blades. The NACA series symmetrical airfoils are often adopted for blades.

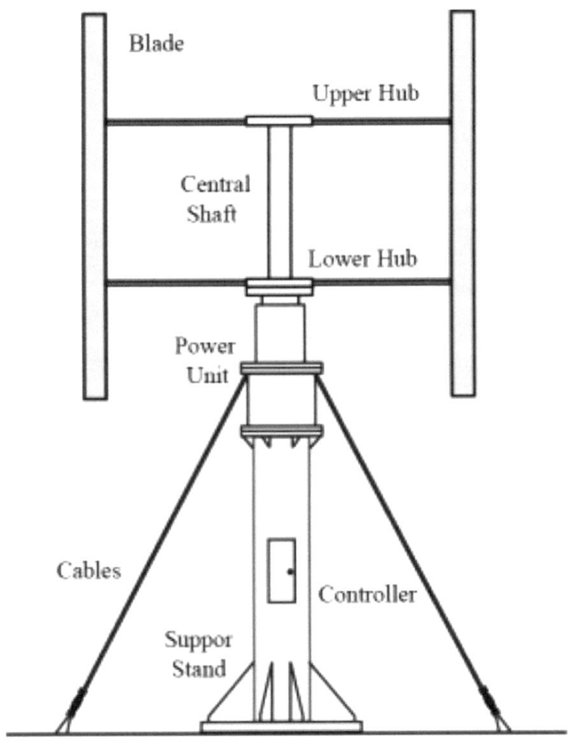

Figure 4.
Straight-bladed VAWT.

3. Aerodynamic characteristics: research methods

3.1 Aerodynamic computational methods

As a kind of wind turbine, the research methods of aerodynamic characteristics for wind turbines are all suitable for SB-VAWT. There are two main kinds of research ways including theoretical method and experimental method. The theoretical method includes aerodynamic computational models and numerical simulation [5, 6]. The experimental method mainly includes wind tunnel test, visualization test, and field test.

Although the structure of SB-VAWT is relatively simple comparing with HAWT, the aerodynamic characteristics are rather complex. **Figure 5** shows the main theoretical methods which are divided into two main categories. There are three typically aerodynamic models including momentum model, vortex model, and cascade model [6]. The first two methods are frequently used. These models are very important for deducing optimum design parameters and also for predicting the performance of SB-VAWT. The momentum theory is the most important method for analyzing the aerodynamic characteristics of blade both for HAWT and VAWT. It has three kinds of theory including momentum model, blade element model, and blade element momentum (BEM) model. The BEM model is the most widely used method even now. Based on the BEM model, an aerodynamic computational model, streamtube model, just used for Darrieus-type VAWT, was proposed. In 1974, Templin [7] proposed the single streamtube model firstly for the aerodynamic performance of Darrieus VAWT. This method is the simplest streamtube model. In the same year, an improved model, multiple streamtube model, was proposed by Wilson and Lissaman [8]. From then on, some other streamtube models were proposed [9–11]. The most important streamtube model, double-multiple streamtube model, was proposed by Ion Paraschivoiu in 1981 [12]. This model is one of the most widely used methods for analyzing the aerodynamic characteristics of Darrieus-type VAWT. The detailed information can be found in the book written by Ion Paraschivoiu [3]. The schematics

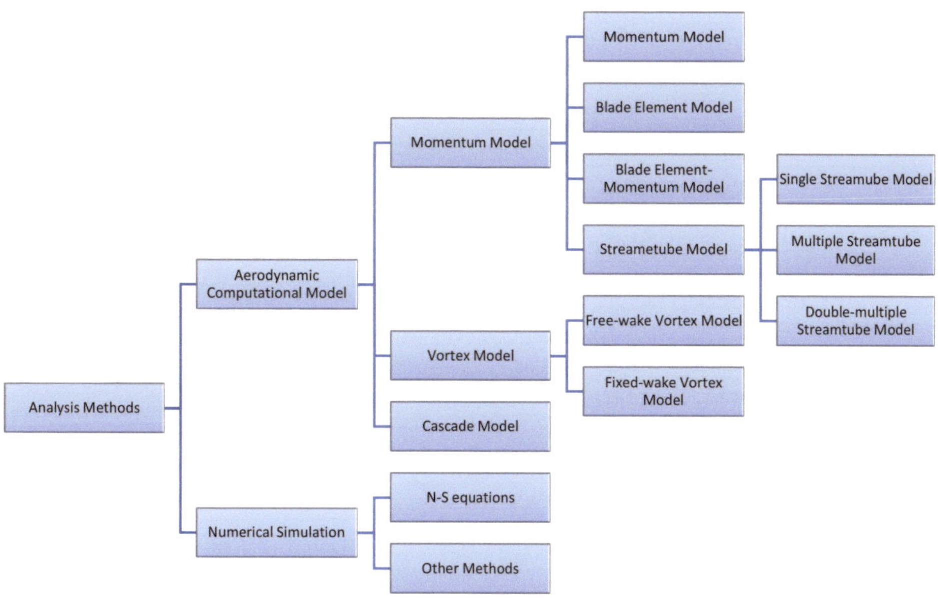

Figure 5.
Main theoretical methods for SB-VAWT.

of three kinds of streamtube model are shown in **Figure 6**. In **Figure 6(a)**, V is the wind velocity, Va is the induced velocity, and Vw is the wake velocity in downstream side. For **Figure 6(b)**, the V1, V2, and V2 are the wind velocities in different stream tubes; the Va1, Va2, and Va3 are the induced velocities in different stream tubes; and Vw1, Vw2, and Vw3 are the wake velocities in different stream tubes.

The streamtube models equate the forces on the rotor blades to a change in streamwise momentum through the wind turbine rotor. Therefore, the momentum model can predict well the overall performance of the wind turbine when the loads on rotor blades are light and the tip-speed ratios are relatively low. However, for the condition of high tip-speed ratios and high rotor solidities, the momentum model is inadequate. Therefore, another aerodynamic model, the vortex model, was proposed and developed based on vorticity equations [13]. The vortex models

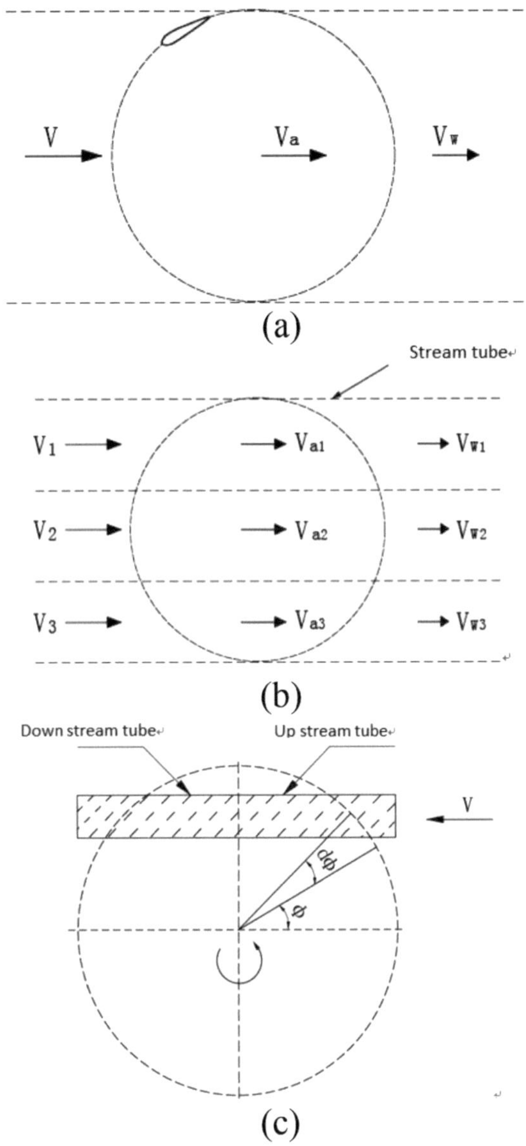

Figure 6.
Schematic of three kinds of streamtube models [6]. (a)Single Streamtube Model, (b) Multiple Streamtube model, (c) Double-Multiple Streamtube model.

are basically potential flow models based on the calculation of the velocity field of the turbine through the influence of vorticity in the wake of the blades. The vortex model includes two kinds, free wake model and fixed wake model. The detailed information can be found in the relative references [6].

Numerical simulation based on computational fluid dynamics (CFD) is another important research method developed rapidly in recent years. With the fast development of computer science and technology, numerical simulation has become a main method for analysis and design of SB-VAWT. More and more researchers begin to use this way to develop a new type of SB-VAWT. In the modern time, the researchers have tried to combine these models together to simulate the aerodynamic performance of SB-VAWT [14–17]. Because the blade of SB-VAWT is straight, the blade section along span direction is the same. Therefore, two-dimensional simulation is often used. Of course, it can be thought that the three-dimensional simulation may have better accuracy. The time and cost for simulation will be higher on the contrary. **Figure 7** shows an example of two-dimensional meshing around the blades of a three-bladed SB-VAWT and the 3D flow field around rotor.

3.2 Experimental methods

Experimental methods can be thought as the most direct and effective method if the tests are well done. Usually, for wind turbine aerodynamic performance, the experiments can be categorized as wind tunnel test, visualization test, and field test. The following introduction of experimental methods is based on the author's research works in Northeast Agricultural University (NEAU) of China [18–20].

3.2.1 Wind tunnel test

Usually, the wind tunnel can be categorized as closed layout type and open jet type. Both of them can be used for aerodynamic characteristics test of wind turbines. Furthermore, according to the section shape of test part, it can be divided into round shape section and square shape section. For SB-VAWT, the wind tunnel with square shape section can be thought as more suitable. **Figure 8** shows a wind tunnel experimental system in NEAU. Considering the blockage effect, blockage ration defined as ration of the swept area of rotor against section area of wind tunnel should be less than 35%. Usually, the power can be obtained by measuring the torque and revolution of rotor. The aerodynamic performance evaluation indicators include power coefficient (Cp), torque coefficient (Ct), and tip-speed ratio (λ) defined as the equations shown below.

$$C_t = \frac{T}{\frac{1}{2}\rho ARU^2} \tag{1}$$

$$C_p = \frac{T\omega}{\frac{1}{2}\rho AU^3} = \frac{P}{\frac{1}{2}\rho AU^3} \tag{2}$$

$$\lambda = TSR = \frac{\omega R}{U} = \frac{\pi Dn}{30U} \tag{3}$$

Where ρ (kg/m3) is air density, U (m/s) is the incoming flow velocity, R (m) is rotor radius, D (m) is rotor diameter, A (m2) is the cross-sectional area of the rotor, ω (rad/s) is angular velocity of the rotor, n (RPM) is the rotational speed of rotor, T (N•m) is the rotor torque, and P (m/s) is output power (W).

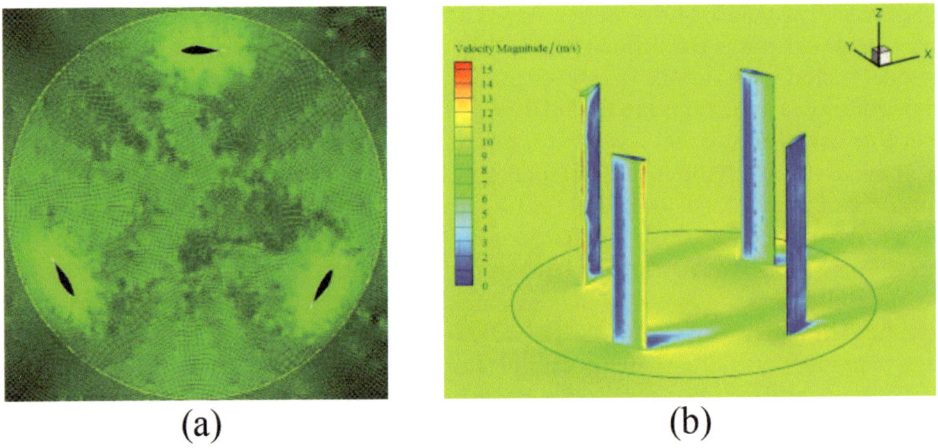

Figure 7.
Examples of numerical simulations for SB-VAWT. (a) 2D meshing and (b) 3D flow field.

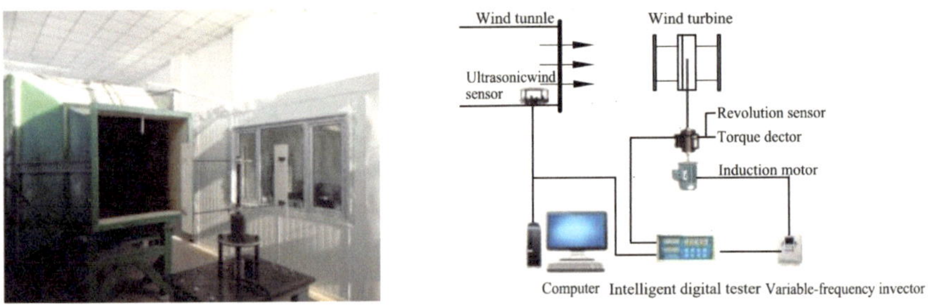

Figure 8.
Wind tunnel test system in NEAU for SB-VAWT.

3.2.2 Visualization test

The flow fields around wind turbine rotor are very complex. To investigate the performance, the flow mechanism should be researched. The visualization tests on flow fields around rotor are an important way. Benefiting from the visual technology, many methods were proposed, such as smoke-wire method, laser method, PIV, PTV, etc.

Figure 9 shows a schematic diagram of the visual experiment system in NEAU by smoke-wire way and the photo of flow field around a three-bladed SB-VAWT rotor obtained in the test. The test model was placed at the same center of the wind tunnel outlet and downstream from the outlet. The smoke-wire generator was set up at the center of the outlet of wind tunnel, and its volt was controlled to generate smoke wire. A CCD camera was placed above the test model to take photo. Visual photos of the flow path lines in and around the rotor were obtained and can be processed by image software in computer.

Recently, the particle image velocimetry (PIV) technology becomes more and more popular. PIV is an optical method of flow visualization used to obtain instantaneous velocity measurements and related properties in fluids. The sufficiently small particles named tracer particles are input into fluid, and their motion can be captured by CCD camera. Based on the relative function and computational methods, the velocity field of the flow can be obtained. Furthermore, the vorticity and pressure distributions around wind turbine blades can be also calculated. **Figure 10** shows an example of PIV system experimental system in NEAU and the flow fields around blade of SB-VAWTs at different azimuth angles.

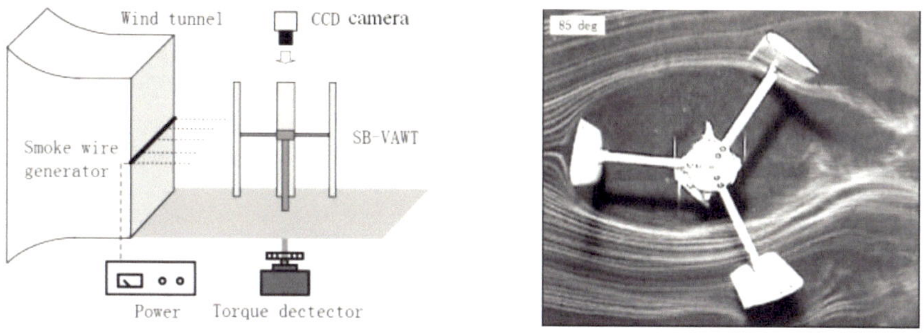

Figure 9.
Schematic diagram of the visual experiment system.

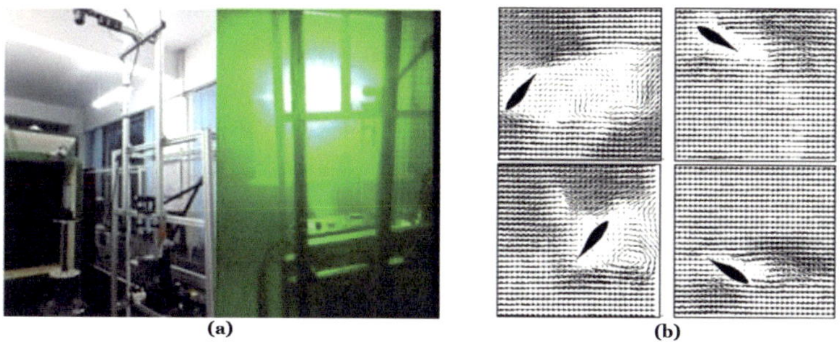

Figure 10.
An example of PIV experimental system for VAWT. (a) PIV system for VAWT and (b) flow fields obtained by PIV test.

Figure 11.
A prototype SB-VAWT for field test in NEAU.

3.2.3 Field test

To obtain the true and reliable aerodynamic performance of VAWT, sometimes, the field tests will be carried out after theoretical research or even wind tunnel tests. A prototype wind turbine, not the rotor only, should be designed and fabricated. The turbine should be located in the area with good wind speed conditions. Continuous power performance measuring and testing are required for a long period of time. **Figure 11** shows a prototype SB-VAWT for field test in NEAU. Because the wind in nature is always changed both the wind speed and direction. The situation is different with the lab. Therefore, the results of field tests are often lower than that obtained by lab test.

4. Aerodynamic performance improvement researches

Since the SB-VAWT was proposed, many efforts have been made to study the ways on its performance improvement. The aerodynamic performance of SB-VAWT has gradually become more and more clear. However, there are still some disadvantages that should be improved comparing with the propeller-type HAWTs. The power performance of SB-VAWT is rather lower than the propeller-type HAWT, and the starting performance under low wind speed is not so good. To improve these disadvantages, investigations have been carried out, and the research methods can be summarized as mainly three ways [21]:

1. By developing new rotor structure or blade airfoil

2. By combining drag-type VAWT rotor

3. By adding additional device

4.1 Blade and rotor structure researches

In the early stage, there was no special airfoil for the blade of SB-VAWT. The blade airfoil for aircraft or HAWT was often adopted, such as NACA series. Prof. Seki from Tokai University in Japan took up the research on blade airfoils for SB-VAWT from 1976 year [22]. **Figure 12** shows a specialized TWT airfoil for SB-VAWT which was exploited in Tokai University. In recent years, more and more airfoils have been researched for the utilization of SB-VAWT. **Figure 13** shows the aerodynamic performance researches of SB-VAWT with 20 kinds of airfoil [23].

Figure 14 shows the main rotor structure types of SB-VAWT. The (a), (b), and (c) in the figure are the most basic types. The others are the improved types for increasing some performance, starting performance, decreasing blade tip loss and rotor structure strength performance, etc. As an example, to tilt the blade at certain angles shown in **Figure 14(d)–(f)** may improve the starting performance. By making the blade bend at certain angles, the loss of blade tip may be decreased as shown in **Figure 14(g)** and **(h)**. For the rotor with long blade, dividing one blade to several stages can reduce blade manufacture cost.

4.2 Combining drag-type VAWT rotor

For the researches on combining with drag-type rotor, as the most famous drag-type VAWT, Savonius rotor is often used to be added in the rotor of SB-VAWT

Figure 12.
TWT airfoil [22].

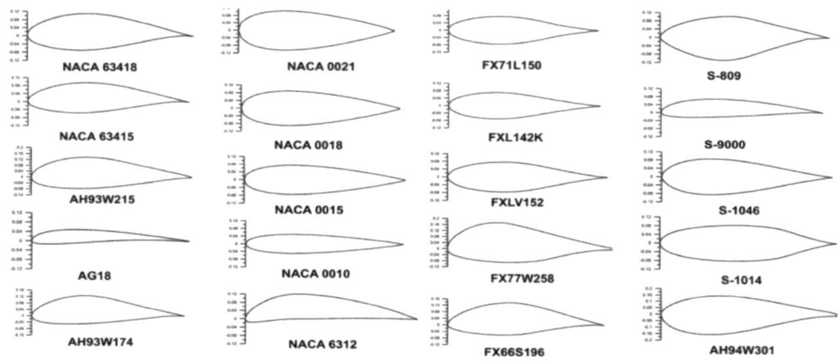

Figure 13.
Twenty kinds of airfoil research for SB-VAWT [23].

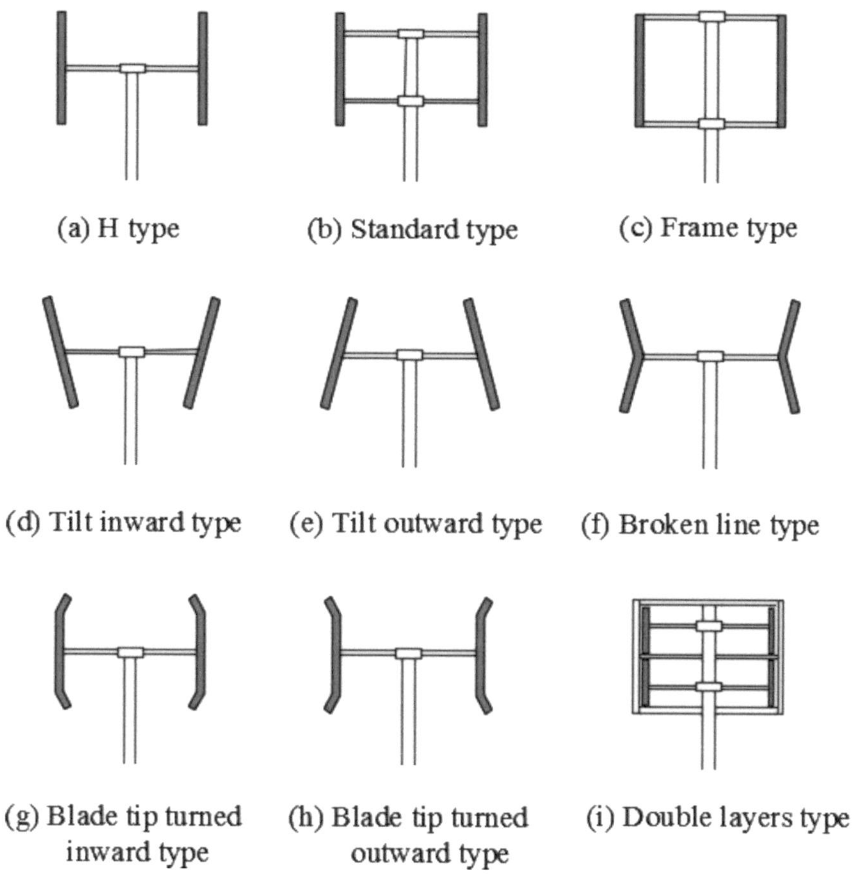

Figure 14.
Main rotor structures of SB-VAWT.

to increase its starting performance [21]. This method is firstly used for increasing the starting performance of Darrieus VAWT because the Darrieus VAWT can start rotation itself as shown in **Figure 15** [24]. **Figure 16** shows some examples of this kind of lift-drag combined-type VAWTs.

Although the improvement of starting performance by adding Savonius rotor is obvious, the power performance at high rotational speed is greatly affected. The reason is mainly because the Savonius rotor will be turned into a load when its

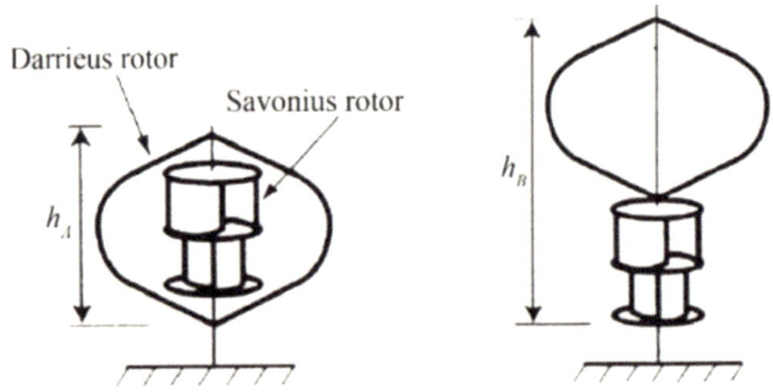

Figure 15.
Savonius-Darrieus combined-type VAWT [24].

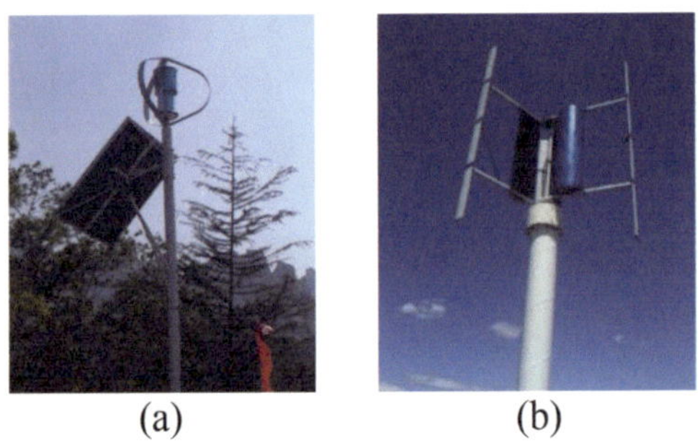

Figure 16.
Lift-drag combined-type VAWTs. (a) Street lamp and (b) wind turbine.

Figure 17.
SB-VAWT with WGD up and down of the rotor.

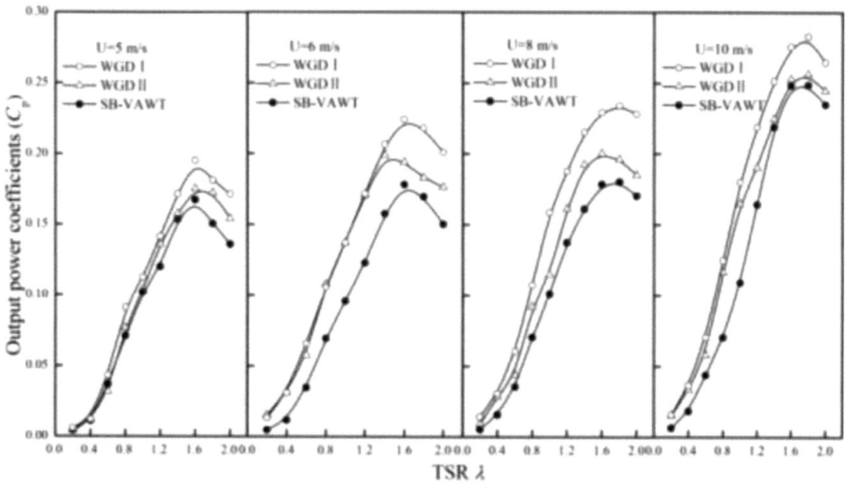

Figure 18.
Cp of SB-VAWT with and without WGD by wind tunnel test [21].

tip-speed ratio becomes larger than unit. Therefore, the combination factors should be deeply researched such as their rotor diameter, combine angle, aspect ratio, etc.

4.3 Adding wind-gathering device

Some researches were carried out on setting additional devices around or up and down the rotor of SB-VAWT to increase its performance. These devices can be called as wind-gathering device (WGD). For the researches on adding WGD, most of researchers set up some different kinds of guide vanes around the rotor along with the wind direction [25]. Another way is to set the WGD up and down the rotor as shown in **Figure 17**. This type of WGD was proposed by Prof. Li from NEAU of China [21]. It is designed as a truncated cone-shaped structure installed up and down of the rotor. The main advantage of this WGD is that it can increase the wind speed from all wind directions. It will not need to take space along the wind direction so that the whole rotor structure will not be suffering from huge thrust and drag force. Furthermore, the WGD can play a certain role in preventing rain, snow, and ice on blades. According to the results of numerical simulation and wind tunnel tests, the power coefficients of the rotor with this WGD were much higher than the rotor without WGD (**Figure 18**).

5. Applications

Wind power generation is mainly including grid-connected and off-grid wind power generation. The large-scale wind turbine is often used for grid-connected power generation. The middle- and small-scale wind turbines are usually for off-grid power generation. For SB-VAWT, at the current stage, it is mainly used for small-scale off-grid wind power generation [26, 27].

5.1 Electrical generation

Figure 19 shows an example of off-grid electrical generation system based on SB-VAWT. The electrical generation system mainly includes SB-VAWT, control device, storage battery, and inverter. It is usually used in the power supply of street lamp, home lighting, etc.

Because the wind in nature is not stable, the wind energy is often used together with other kinds of renewable energy, such as solar energy, biomass energy, water energy, etc. This type of energy system is called hybrid energy system. The SB-VAWT also can be used as this type system. The most popular type of hybrid energy system for SB-VAWT is to be used together with solar energy. **Figure 20** shows the schematic diagram of an example of this system.

5.2 Other applications

There are many other applications for SB-VAWT excepted for electrical power generation. From the point of energy conversion, wind power can be transferred to

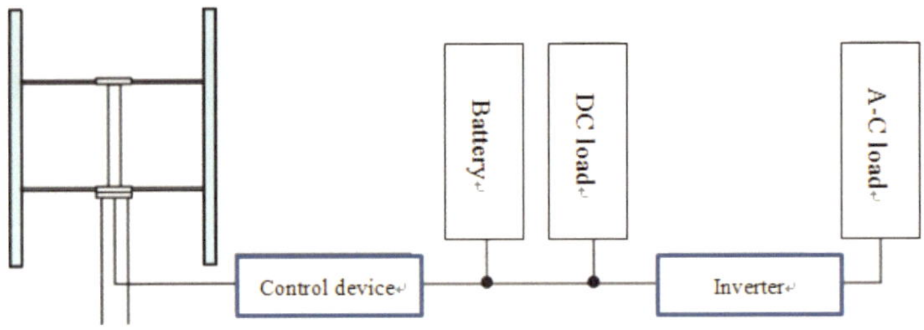

Figure 19.
Schematic diagram of an electrical generation system with SB-VAWT.

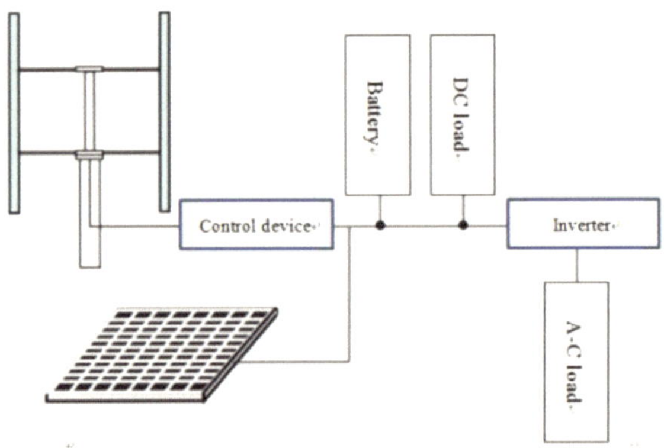

Figure 20.
Solar wind hybrid energy system.

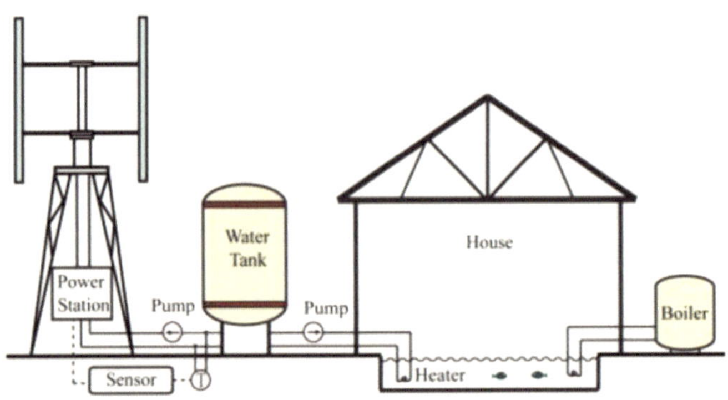

Figure 21.
Concept of fishpond heating system by SB-VAWT.

mechanical energy and heat energy. Therefore, the SB-VAWT energy system can be also used for water pumping, heating or cooling, etc. **Figure 21** shows a concept of fishpond heating system by wind power with SB-VAWT.

6. Summary

More and more attentions have been paid to the research on straight-bladed vertical axis wind. Many methods have been proposed for improving the aerodynamic performance. As the rotor becomes larger and larger, the utilization fields of SB-VAWT will become wider and wider in the future.

Acknowledgements

This word is sponsored by the Project 51576037 supported by National Natural Science Foundation of China (NSFC). The authors would like to give thanks to their supporters.

Author details

Yan Li
Engineering College, Northeast Agricultural University, Harbin, China,

*Address all correspondence to: liyanneau@163.com

IntechOpen

© 2019 The Author(s). Licensee IntechOpen. This chapter is distributed under the terms of the Creative Commons Attribution License (http://creativecommons.org/licenses/by/3.0), which permits unrestricted use, distribution, and reproduction in any medium, provided the original work is properly cited. CC BY

References

[1] Li Y. The vertical axis wind turbine technology lecture one—The vertical axis wind turbine and its development situation. Renewable Energy Sources. 2009;**27**(1):121-123. (in Chinese)

[2] Li Y, Zheng Y, Zhao S, et al. A review on aerodynamic characteristics of straight-bladed vertical axis wind turbine. Acta Aerodynamica Sinica. 2017;**35**(03):368-382. (in Chinese)

[3] Paraschivoiu I. Wind Turbine Design with Emphasis on Darrieus Concept. Montreal, Quebec, Canada: Polytechnic International Press; 2002

[4] Darreieus G. Turbine having its rotating shaft transverse to the flow of current. U.S. Patent 1834018. December 1931

[5] Jin X, Zhao G, Gao KJ, et al. Darrieus vertical axis wind turbine: Basic research methods. Renewable and Sustainable Energy Reviews. 2015;**42**:212-225

[6] Islam M, David S, Ting K, et al. Aerodynamic models for Darrieus-type straight-bladed vertical axis wind turbines. Renewable and Sustainable Energy Reviews. 2008;**12**:1087-1109

[7] Templin RJ. Aerodynamic performance theory for the NRC vertical-axis wind turbine. NRC Lab. Report LTR-LA-190; June 1974. p. 29

[8] Wilson RE, Lissaman PBS. Applied Aerodynamics of Wind Power Machines. Corvallis, OR, USA: Oregon State University; 1974

[9] Strickland JH. A performance prediction model for the Darrieus turbine. In: International Symposium on Wind Energy Systems; 7-9 September 1976; Cambridge, UK. pp. C3-39-54

[10] Mureaca RJ, Stephens MV, Dagenhart JR. Theoretical Performance of Cross-Wind Axis Turbines with Results for a Catenary Vertical Axis Configuration. NASA TMX-72662. USA; 1975

[11] Sharpe DJ. A Theoretical and experimental study of the Darrieus vertical axis wind turbine. Research Report. School of Mechanical, Aeronautical & Production Engineering. Kingston Polytechnic; October 1977

[12] Paraschivoiu I. Double-multiple streamtube model for Darrieus wind turbines. Second DOE/NASA Wind Turbines Dynamics Workshop. NASA CP-2186, Cleveland, OH, February; 1981. pp. 19-25

[13] Wilson RE. Vortex sheet analysis of the Giromill. ASME Journal of Fluids Engineering. 1978;**100**:340-342

[14] Zhenzhou Z, Honghong Y, Yuan Z, et al. Numerical simulation on unsteady flow field of straight bladed Darrieus wind rotor. Journal of Drainage and Irrigation Machinery Engineering. 2012;**30**(5):567-572. (in Chinese)

[15] Li Y, Liu QD, Wang SL, et al. Wind tunnel test and numerical simulation on blade icing of small scale vertical axis wind turbine. Acta Aerodynamica Sinica. 2016;**34**(5):568-572,586. (in Chinese)

[16] Zhao Z, Qian S, Zheng Y, et al. Enhancement approaches of aerodynamics performance of lift type vertical axis wind turbine considering small angle of attack. Journal of Drainage and Irrigation Machinery Engineering. 2018;**36**(2):146-453. (in Chinese)

[17] Wenju J, Jianwen W, et al. Analysis of vertical axis wind turbine power changing with rotating speed from vertical and horizontal flow fields. Journal of Drainage and

Irrigation Machinery Engineering. 2018;**36**(2):166-171. (in Chinese)

[18] Kotaro T, Yan L. A wind tunnel experiment of self-starting capability for straight bladed vertical axis wind turbine. Journal of Drainage and Irrigation Machinery Engineering. 2018;**36**(2):136-140

[19] Feng F, Zhao SY, Qu CM, et al. Research on aerodynamic characteristics of straight-bladed vertical axis wind turbine with S series airfoils. International Journal of Rotating Machinery. 2018;**2018**:13

[20] Li Y, Zhao S, et al. PIV visualization experiment on static flow field of Savonius wind turbine. Journal of Drainage and Irrigation Machinery Engineering. 2018;**36**(2):159-165. (in Chinese)

[21] Li Y, Zhao SY, Kotaro T, et al. Starting performance effect of a truncated-cone-shaped wind gathering device on small-scale straight-bladed vertical axis wind turbine. Energy Conversion and Management. 2018;**167**(1):70-80

[22] Seki K. Vertical axis wind turbine designed aerodynamically at Tokai University. Periodica Polytechnica. 1981;**25**(1):47-56

[23] Mohamed MH. Performance investigation of H-rotor Darrieus turbine with new airfoil shapes. Energy. 2012;**47**:522-530

[24] Tetsuya W, Yoshiaki T, Takumi H, et al. Hybrid configuration of Darrieus and Savonius rotors for stand-alone wind turbine-generator systems. Electrical Engineering in Japan. 2005;**150**(4):13-22

[25] Chong WT, Fazlizan A, Poh SC. The design simulation and testing of an urban vertical axis wind turbine with the omni-direction guide-vane. Applied Energy. 2013;**112**(SI):601-609

[26] Hui I, Cain BE, Dabiri JO. Public receptiveness of vertical axis wind turbines. Energy Policy. 2018;**112**:258-271

[27] Tjiu W, Marnoto T, Mat S, et al. Darrieus vertical axis wind turbine for power generation. II: Challenges in HAWT and the opportunity of multi-megawatt Darrieus VAWT development. Renewable Energy. 2015;**75**:560-571